SCIENCE FOR EXCELLENCE

physical
science

LEVEL 3

Scottish Schools Science Group

Series Editors:
Nicky Souter, Paul Chambers and Stephen Jeffrey

Authors:
Paul Chambers, Kevin Christie, Tom Clark,
Ross McMahon and Caroline Scott

The front cover shows a computer artwork of electrical sparks crossing between two plasma spheres.

Although every effort has been made to ensure that website addresses are correct at time of going to press, Hodder Gibson cannot be held responsible for the content of any website mentioned in this book. It is sometimes possible to find a relocated web page by typing in the address of the home page for a website in the URL window of your browser.

Hachette UK's policy is to use papers that are natural, renewable and recyclable products and made from wood grown in sustainable forests. The logging and manufacturing processes are expected to conform to the environmental regulations of the country of origin.

Whilst every effort has been made to check the instructions of practical work in this book, it is still the duty and legal obligation of schools to carry out their own risk assessment.

Orders: please contact Bookpoint Ltd, 130 Milton Park, Abingdon, Oxon OX14 4SB. Telephone: (44) 01235 827720. Fax: (44) 01235 400454. Lines are open 9.00–5.00, Monday to Saturday, with a 24-hour message answering service. Visit our website at www.hoddereducation.co.uk. Hodder Gibson can be contacted direct on: Tel: 0141 848 1609; Fax: 0141 889 6315; email: hoddergibson@hodder. co.uk

© Scottish Schools Science Group 2010
First published in 2010 by
Hodder Gibson, an imprint of Hodder Education,
An Hachette UK Company
2a Christie Street
Paisley PA1 1NB

Impression number   5 4 3 2 1
Year   2012 2011 2010

All rights reserved. Apart from any use permitted under UK copyright law, no part of this publication may be reproduced or transmitted in any form or by any means, electronic or mechanical, including photocopying and recording, or held within any information storage and retrieval system, without permission in writing from the publisher or under licence from the Copyright Licensing Agency Limited. Further details of such licences (for reprographic reproduction) may be obtained from the Copyright Licensing Agency Limited, Saffron House, 6–10 Kirby Street, London EC1N 8TS.

Cover photo PASIEKA/SCIENCE PHOTO LIBRARY
Illustrations by Emma Golley at Redmoor Design, Tony Wilkins, and DC Graphic Design Limited
Typeset in Minion 12/15pt by DC Graphic Design Limited, Swanley, Kent
Printed in Italy

A catalogue record for this title is available from the British Library

ISBN: 978 1444 110 760

Contents

Introduction

The first three Science for Excellence titles support learning at the Level Three outcomes of Curriculum for Excellence. This title focuses on Forces, Electricity and Waves and draws from the other organisers. It attempts to form a coherent link between the second and third level outcomes. It also makes frequent links to the key concepts identified in Curriculum for Excellence in that the chapters' contents cross into the organisers Planet Earth, Biological Systems, Materials and Topical Science.

Where appropriate, the Science for Excellence titles use familiar content and approaches while also embracing the principles of Curriculum for Excellence. The books also attempt to take the topics through to a natural conclusion and to provide, where appropriate, more demanding contexts for pupils.

Modern applications feature prominently throughout the chapters. The book provides many real life examples and applications of the principles which create opportunities for pupils to learn and appreciate the factors which led to the scientific discoveries while also being aware of the implications of scientific advances and their impact on society and the environment. It is hoped that extending the scope of the content beyond the traditional 'physics' boundaries will develop a more rounded appreciation of science and society and will lead to greater motivation and a deeper understanding of the issues.

Some of the activities in the book involve experiments. These should only be attempted under the instruction of the Science Teacher and in accordance with the appropriate safety guidelines. Questions and activities are designed to examine and extend the content of the chapters. Skills in literacy and numeracy as well as an awareness of the importance of health and wellbeing will be developed through these exercises – look out for the icons shown at the end of this Introduction. Some chapters allow for numerical and graphical activities while others seek to reinforce the scientific principles contained in the main text. Curriculum for Excellence encourages learners to become active participants and the Active Learning activities in this Science for Excellence series encourage open-ended and pupil investigation activities as well as individual and group project and research work where learners are asked to make informed decisions on scientific advances which may have ethical or societal implications. Tasks are designed around the 'broad features of assessment in science'.

By engaging in the activities and tasks pupils will show features of the skills sought in the 'principles and practice' documentation. Pupils will have the opportunity to demonstrate work that will allow teachers to assess:

- How well do they contribute to investigations and experiments?

- Are they developing the capacity to engage with and complete tasks and assignments?

- To what extent do they recognise the impact the sciences make on their lives, on the lives of others, on the environment and on society?

The principles and practices outlined in Curriculum for Excellence have been adopted throughout Science for Excellence. The series is designed to be used in conjunction with schemes of work which reflect learning and teaching approaches which are most applicable to the sciences. The chapters provide opportunities for scientific enquiry and examples of scientific scenarios where pupils can, for example, link variables to determine relationships, improve their scientific thinking or make informed judgements on the basis of scientific principles.

Scientifically Literate Citizens

The use of real data and experimental-type situations will help to develop scientific attitudes. Pupils will be able to look at the data critically and make informed judgements on the basis of what is in front of them. Additionally, they will be critical of broad or bold claims and be able to analyse the science as well as the implications of such claims. Ultimately, the significant challenge for CfE is that it changes pupils' attitudes to science and makes them more able to engage positively in issues that will affect them; that they are able to understand the scientific challenges and issues facing them and respond in a critical and informed manner.

CfE documentation states that learning in the sciences will enable pupils to:

- develop curiosity and understanding of the environment and my place in the living, material and physical world

- demonstrate a secure knowledge and understanding of the big ideas and concepts of the sciences

- develop skills for learning, life and work

- develop the skills of scientific inquiry and investigation using practical techniques

- develop skills in the accurate use of scientific language, formulae and equations

- apply safety measures and take necessary actions to control risk and hazards

- recognise the impact the sciences make on my life, the lives of others, the environment and on society

- recognise the role of creativity and inventiveness in the development of the sciences

- develop an understanding of the Earth's resources and the need for responsible use of them

- express opinions and make decisions on social, moral, ethical, economic and environmental issues based upon sound understanding

- develop as a scientifically-literate citizen with a lifelong interest in the sciences

- establish the foundation for more advanced learning and future careers in the sciences and the technologies.

 Literacy

Numeracy

Health and wellbeing

FORCES, ELECTRICITY AND WAVES

Waves

1

Electromagnetic waves

Level 2   What came before?

● SCN 2–11b

By exploring reflections, the formation of shadows and the mixing of coloured lights, I can use my knowledge of the properties of light to show how it can be used in a creative way.

Level 3   What is this chapter about?

● SCN 3–11b

By exploring radiations beyond the visible, I can describe a selected application, discussing the advantages and limitations.

Electromagnetic waves

In 1820, Danish scientist Hans Christian Ørsted made an unusual discovery.

Hans Christian Ørsted

He was giving an evening lecture to friends and colleagues in his home. The topics of the lecture were Electricity and Magnetism. He intended to demonstrate, firstly, that passing an electric current through a wire would make it heat up and glow and, secondly, to demonstrate magnetism by using compasses and permanent magnets. During the first experiment, however, Ørsted noticed that the needles of his compasses moved when he switched the current on and off. He kept quiet about what he saw, and the next day started to try to work out why this had happened. Over the next few months he worked on finding the reason behind this movement of the compass needles. However, the reason eluded him, and Ørsted published his findings in 1820 without an explanation of this curious effect.

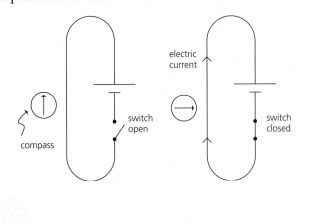

electric current

switch open

compass

switch closed

Scientists everywhere were immediately intrigued by Ørsted's findings, none more so than the French mathematician and scientist André-Marie Ampère.

André Marie Ampère

Within a few weeks, he quickly designed more experiments. In one, he wrapped a wire around an iron rod, then passed a current through the wire, turning the iron rod into a magnet (or as we would now call it, an electromagnet). Ampère delved deeper into this strange effect and managed to produce a *mathematical formula* which helped describe the relationship between electric current and magnetism.

This formula is beyond your understanding at the moment ($\phi B.ds = \mu_0 I$)!

Shortly after this, in 1821, the English physicist Michael Faraday was also similarly intrigued by Ørsted's publication, and he started to research the phenomena for himself.

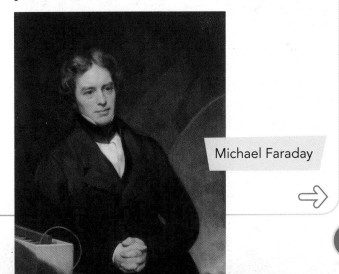

Michael Faraday

Michael Faraday wasn't the greatest mathematician, but he was incredibly good at experimental work. His diagrams of the experiments and how they operated were detailed and showed the results very well. He wasn't keen on Ampère's work, or his hypothesis that magnetism was caused by moving charged particles. Faraday designed new experiments and these showed that electricity and magnetism were linked somehow, and that they could produce movement. He also did work on something called Electromagnetic Induction and continued to work on electromagnetism, off and on, throughout the rest of his life. In 1846 he gave a lecture, called 'Thoughts on ray-vibrations', suggesting that magnetism can affect light.

The experimental equipment used by Michael Faraday

There is no doubt that Faraday was a genius who contributed hugely to science and education. Among other things, he started the Royal Institute Christmas Lectures for children, and with a total of nineteen, still holds the record for the number of lectures given. In fact Albert Einstein had a photograph of him on his study wall for inspiration. Two other eminent physicists also had their portraits on Einstein's wall. One of them was Sir Isaac Newton, and the other was the Scottish physicist James Clerk Maxwell.

Faraday was a great, if not the greatest, experimental scientist. He was largely self-taught and very practical. His real skill lay in an ability to design experiments which showed which particular factor could cause other effects.

James Clerk Maxwell

James Clerk Maxwell, on the other hand, was able to take Faraday's work and model it mathematically. He did this with the work of others who had worked on electromagnetism, and he produced a set of four equations, called the **Maxwell Equations**, which describe beautifully electric and magnetic waves. Maxwell's work paved the way for Einstein to develop his Special Theory of Relativity!

Sadly, James Clerk Maxwell is not as famous as he should be. He attended school in Edinburgh and was one of the finest mathematicians of his time. He held many posts at some of the most prestigious universities in Britain and contributed to wide range of subjects. He died at the early age of 48 and a statue in his honour was unveiled in Edinburgh in 2008.

This statue of James Clerk Maxwell's stands in the New Town area of Edinburgh

He is buried in Parton Kirk cemetery near Castle Douglas. Einstein could not have done what he did without Maxwell's work. Albert Einstein knew exactly how important Maxwell's equations were and said that Maxwell's work was

'the most profound and the most fruitful that physics has experienced since the time of Newton'.

So how does this tie in with electricity and magnetism? One strange result of Maxwell's equations was that they unexpectedly involved the speed of light.

James Clerk Maxwell did many things, from explaining the cause of Saturn's Rings to producing the first colour photograph, but his greatest achievement was in electromagnetism, and fundamental to that was his realisation that light was a wave and this was composed of electrical and magnetic 'vibrations'.

As well as appreciating that light was an electromagnetic wave, James Clerk Maxwell realised this meant that visible light was just one part of a wide range of electromagnetic waves, and that there were almost certainly other types of electromagnetic waves still to be discovered.

Later, in 1887, Heinrich Hertz proved by experiment that light was indeed an electromagnetic wave. Guigliemo Marconi, Nikola Tesla and other pioneers exploited the phenomenon commercially, and this understanding of the electromagnetic spectrum has led to our use of radio, TV, mobile phones and satellite communication.

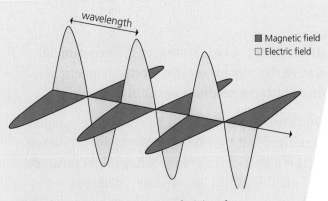

wavelength

■ Magnetic field
☐ Electric field

The electric and magnetic fields of an electromagnetic wave

The clues had been there from when Faraday showed experimentally that magnetism can affect light. He had in fact been using magnetism to affect an electromagnetic wave.

Active Learning ▶

Activities

Prepare a poster or presentation on one of these famous scientists already mentioned in this chapter:

James Clerk Maxwell; Michael Faraday; Heinrich Hertz; Guigliemo Marconi; Nikolai Tesla.

Your poster/presentation has to include one image of the scientist and one image of an experiment or piece of apparatus he used. You have to provide details of his birth, education, discoveries and death. Any other information you consider relevant may be included.

The poster can comprise of two pages of A4 or 4 PowerPoint slides if that is possible.

The production of electromagnetic waves

When we talk about the **Electromagnetic Spectrum**, we mean the whole range of electromagnetic waves, from long, weak radio waves at one end to short, high energy gamma rays at the other end.

Visible light, the light that our eyes detect and we use to see, is just one kind of electromagnetic wave. We should however be familiar with the answers to the following questions:

- What are electromagnetic (EM) waves?

- How do they travel?

- How are EM waves produced?

What are electromagnetic waves?

The first question is difficult to answer at this level and many physicists struggle to answer it in simple, straightforward terms. You need to know a lot more about Physics to be able to fully understand what an EM wave is. Even then, different physicists interpret the waves in different ways. Let's just settle on the wave being a combination of electric and magnetic fields. These 'fields' are regions where electric and magnetic forces are effective. Hmmm…

A radio transmitter

How do they travel?

To answer the second question, let's start with an example. Water waves use water as a **medium**. This means that water waves need water to travel through. Some kind of energy creates a disturbance in the water, and this energy travels through the liquid in the form of a wave.

Similarly, sound waves create a disturbance in air **particles** and the air particles transmit the energy. So we can say that sound waves use air as the medium they travel through.

When James Clerk Maxwell turned his thoughts to light, he felt there was an invisible medium that light waves used to travel through. He called it the 'luminiferous aether'.

Experiments were done to find and measure this 'luminiferous aether' but none managed to find it. It took another genius, Albert Einstein, to realise

that it doesn't exist! Strangely, the experiments lead to the conclusion that electromagnetic waves **do not need a medium** to travel through!

Don't worry, for light can be even more complex than this! Just try asking your physics teachers if light is a particle or a wave. They might mumble and say things like 'sometimes', 'depends.'

How are EM waves produced?

Well, all waves are caused by vibrations.

Good Vibrations

If you have a big tank of water, and you use energy to move your hand back and forth in the water, you create water waves.

If you speak or shout or sing, you use energy to make your vocal chords vibrate and cause waves in the air particles in your throat and mouth.

$\Rightarrow$

The production of electromagnetic waves

Similarly, if you are a musician you use energy to create vibrations by

- plucking a string of a guitar or other string instrument
- bowing a string of a violin or other string instrument
- making the reeds vibrate in the clarinet, saxophone and other reed instruments
- making the air vibrate in a trumpet, trombone, and other brass instruments
- making the drumskin vibrate on a drum, or the metal vibrate in a cymbal.

Electromagnetic waves are somewhat similar. We don't move water, air or metal to produce EM waves. Instead we move electrically **charged particles**.

Electromagnetic waves are produced when we make the charged particles change speed. They are made to accelerate.

The easiest way to do this is to use an electric circuit!

If we take an electric circuit and switch the battery on and off, we are making the charged particles (electrons) change speed (accelerate).

A better way to do this is to use what is called an alternating current or a.c. curcuit. When we use a.c., the electrons are made to move **back and forth** in the wire. It's like you are constantly changing the battery around, making the electrons flow one way then the other.

Alternating current power supplies do this, but do it really quickly (the mains voltage does it 50 times per second!).

By making electrons move back and forth (oscillate) in a wire in this way we are making them accelerate and so we can produce EM waves.

Types of electromagnetic waves

The hidden rainbow

The **visible light spectrum** is just one tiny part of a much larger range of waves. Sir Isaac Newton showed that if we shine a beam of white light through a glass prism, we can split it up into different colours. This is what happens when light shines through raindrops in the atmosphere to make a rainbow.

This shows that **white light** is made up of different colours of light!

But what makes one colour different from another? What is the difference between blue and red? The answer has to do with the **frequency** of the light.

$\Rightarrow$

Frequency is the term used in Physics to describe how many waves are produced every second.

If one complete wave is produced every second, then the frequency is one **Hertz**, or 1 Hz (for short). This unit of frequency is named after Heinrich Hertz. (Remember him?)

If we produce two complete waves per second, then the frequency is two Hertz, or 2 Hz (for short).

However, if we produce only half a wave every second then we have a frequency of 0.5 Hz. (It takes 2 seconds to produce one complete wave.)

The frequencies of visible light are very high:

- the frequency of red light is roughly 450 000 000 000 000 Hz

- the frequency of blue light is roughly 650 000 000 000 000 Hz.

This means that the little red standby light on your TV is producing 450 000 000 000 000 light waves per second!

The Electromagnetic Spectrum

The complete list of the Electromagnetic Spectrum, in order from lowest to highest frequency, is:

- Radio waves
- Microwaves
- Infrared
- Visible light
- Ultraviolet
- X-rays
- Gamma rays

These radiation groups may be represented by 'bands' in the following diagram.

So what exactly *is* the difference between a light wave and a radio wave?

Well, in some respects, not a lot.

They are both electromagnetic waves and as such they travel at the same speed.

This isn't just any usual speed. It is the fastest speed at which anything can travel. It is 300 million metres per second! This is the speed of light too.

It is usual to refer to the speed of **all** EM waves as the speed of light and not the speed of radio waves, microwaves, or x-rays.

Radio waves, microwaves, infrared, ultraviolet, x-rays and gamma rays are all loosely speaking different kinds of light. We only call visible light 'visible' because our eyes have evolved to see its particular range of frequencies.

The Electromagnetic Spectrum

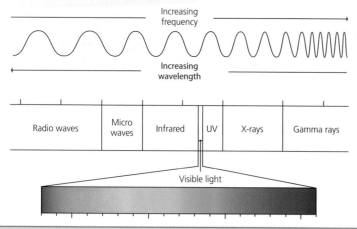

The visible light spectrum

Types of electromagnetic waves

Most insects see in the range of frequencies that we call ultraviolet. They don't see the world as we do. If insects were to start writing Physics textbooks, they could well argue that what we call ultraviolet is actually visible light to them!

Back to the question 'what is the difference between radio waves and light waves?'

Well, although they are both EM waves, and both travel at the speed of light (300 000 000 metres per second), they have different frequencies. To be more accurate, they have different ranges of frequencies.

- Radio frequencies range from just above 0 Hz up to about 300 000 000 000 Hz.

- Visible light frequencies range from roughly 400 000 000 000 000 Hz to 750 000 000 000 000 Hz.

This is not the only difference though.

They both interact with matter differently. For example, radio waves can travel through walls and our bodies, but light cannot. Gamma rays and x-rays at the other end of the spectrum can pass through our bodies but for different reasons from radio waves.

Another way for us to sort the waves in the Electromagnetic Spectrum is to use their **wavelength**. Wavelength is the length of one complete wave, and it is usually measured in metres (m).

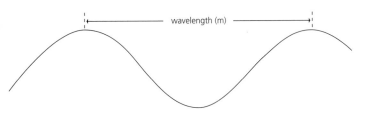

- Radio waves have the longest wavelengths and radio stations broadcast waves with wavelengths from 1 m up to 1 km.

- Gamma rays have the shortest wavelengths and are usually measured in picometres. ('Pico' is a prefix meaning one million millionth. So a wavelength of one picometre is in fact 0.000 000 000 001 m.)

- Different colours of visible light have different wavelengths.

  Red has a wavelength of 700 nanometres, or 0.000 0007 m

  Violet has a wavelength of 400 nanometres, or 0.000 0004 m

('Nano' is another prefix and means one thousand millionth.)

Our eyes have evolved to see using visible light. It is possible however to look at the entire EM spectrum through different 'eyes'.

That's exactly what scientists and engineers have done. They have built detectors which can see these other kinds of light. Evolution has done a fine job of providing us with electromagnetic wave detectors (our eyes). But visible light is just one tiny part of the whole spectrum. If we use other detectors, we can literally see the world, and indeed, the universe around us in a different light.

We can then see things that are normally invisible to our eyes.

Let's have a look then at the different types of electromagnetic waves and what we can do with them.

GLOSSARY

Charged particles particles which have an electrical charge: positive or negative

Electromagnetic Spectrum the range of electromagnetic radiation which includes x-rays, visible light, microwaves and radio waves

Frequency the number of waves in a second

Hertz the unit used to measure frequency

Medium a substance (solid, liquid or gas) through which waves can travel

Particles extremely small amounts of matter

Visible light spectrum the range of colours which our eyes can detect

Wavelength the measurement of the length of a wave of light

FORCES, ELECTRICITY AND WAVES

Waves

2

High frequency electromagnetic waves

Level 2 — What came before?

SCN 2–11b

By exploring reflections, the formation of shadows and the mixing of coloured lights, I can use my knowledge of the properties of light to show how it can be used in a creative way.

Level 3 — What is this chapter about?

SCN 3–11b

By exploring radiations beyond the visible, I can describe a selected application, discussing the advantages and limitations.

High frequency electromagnetic waves

In the first chapter we learned how electromagnetic waves covered a large spectrum and could be described by terms such as 'radio', 'light', and so on depending on frequency and wavelength. The wave bands shown in the diagram on page 12 appear to be distinct, but this is not actually the case. A high frequency UV wave might also be described as a low frequency x-ray.

In this chapter we look at the **high frequency** side of the spectrum and the waves which may be classified as '**gamma rays**', '**x-rays**', and '**UV rays**'.

Gamma rays

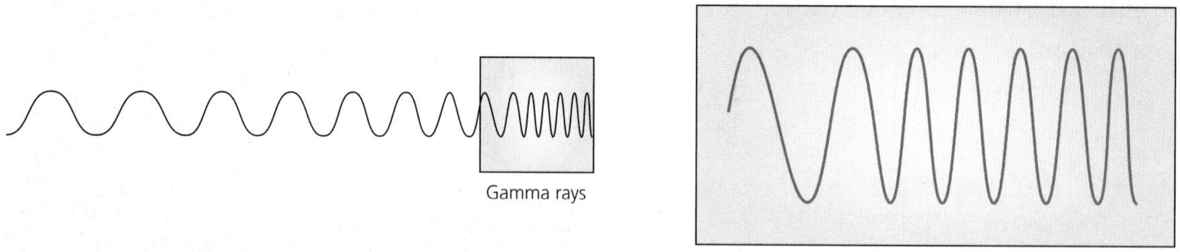

Gamma rays

Gamma rays have the highest frequency and the shortest wavelength

In 1900, the French scientist Paul Ulrich Villard discovered gamma rays when he was working with a radioactive substance called radium. Gamma rays are given out by a number of radioactive materials.

Gamma rays have the shortest wavelengths of all electromagnetic waves. This means they also have the highest frequencies.

Gamma rays are the most energetic of all the types of EM radiation. They are so energetic, that they can pass through all human body tissue and bone. In fact gamma rays can penetrate and pass through materials as dense as steel and concrete!

In the 1960s, during the Cold War, America launched spy satellites to look for nuclear bomb tests by the Soviet Union. The satellites worked by detecting gamma rays given off during the explosions. Scientists soon found gamma rays, but there were two big surprises for them.

- Firstly, the quantity of gamma rays detected was far higher than that from a standard nuclear warhead. In fact, the quantity of gamma rays detected by the satellite was far, far larger than could be produced using ALL the nuclear bombs on Earth!

- Secondly, the gamma rays weren't coming from Russia. They weren't coming from anywhere on Earth at all. They weren't even coming from our galaxy. They were coming from deep space!

Gamma ray detectors are now used to 'look' at the universe and give us information on stars and galaxies. Gamma ray observatories need to be in space, because the rays are absorbed a great deal by our atmosphere.

$\Rightarrow$

Gamma rays

Gamma map from space

Several types of objects in space have been found to emit gamma rays.

1   **Neutron stars**: These are formed when normal stars (slightly larger than our Sun) reach a certain stage in their evolution. When large stars get old, they collapse due to their own gravity. The core becomes so dense that the star cannot collapse any further. The huge pressure causes the star to explode releasing a huge amount of energy. These gamma ray bursts can release more energy in the space of a few seconds than our Sun releases in billions of years! This massive explosion, called a **supernova**, is brighter than an entire galaxy. Interestingly, it is also where the heavy elements from the periodic table are formed. If you are wearing gold or silver jewellery, then the gold or silver atoms were formed in a supernova billions of years ago! As were the atoms that make up your body!

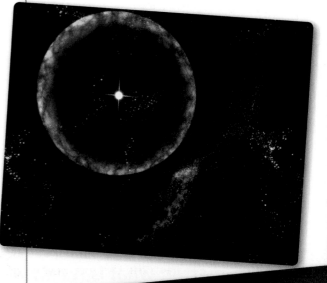

After a supernova explosion, a tiny, incredibly dense object, called a neutron star, is left behind.

Very interestingly, neutron stars give out gamma radiation.

2   A spinning neutron star is called a **pulsar**. A pulsar emits a beam of gamma radiation which 'pulses' because the neutron star is spinning, in rather the same way as a lighthouse sends out light beams.

The pulsar was discovered by Jocelyn Bell Burnell in 1968. Burnell did her first degree at Glasgow University before she moved to the radio astronomy department at Cambridge University. She was a young postgraduate at Cambridge when she made the discovery.

3   There are many types of stars in our universe, many of them much more massive than our Sun. If their mass is more than 20 times the mass of our Sun, when they collapse in on themselves the process keeps on continuing. They keep collapsing in on themselves, becoming so small and dense and that not even light can escape. Space and time are twisted and bent, forming the most mysterious and dangerous object of all in the universe – a **black hole**.

When a black hole forms near another star, the black hole's great gravitational pull will pull in that star. When the star is being ripped apart and is 'falling' into the black hole, huge jets of gamma rays are produced. By searching for these gamma rays we may be able to detect the whereabouts of black holes!

What then was the source of the massive bursts of gamma rays detected by the American spy satellites? It wasn't neutron stars or pulsars. It wasn't the radiation from black holes. Scientists are still not quite sure.

They have linked gamma ray bursts with very old, very distant galaxies, but the true cause is not certain.

Maybe the next generation of physicists – you – might discover the truth!

Anti-matter: what is it and where has it gone?

It is known that radioactive materials emit gamma rays. Physicists also know that gamma rays are given off when matter meets **anti-matter**.

Anti-matter is a very unusual type of matter: It has the **same mass** but **opposite charge** to normal matter. For example, an anti-electron (called a positron) has the same mass as an electron, but instead of possessing a negative charge, it is positively charged.

Our universe is made up of matter, so anti-matter can only be produced under special circumstances, and only exists for a very short time. As soon as an anti-matter particle and a particle of matter collide, the two particles **annihilate**, and are converted into energy in the form of gamma rays.

In 1995, scientists at CERN created and contained, anti-electrons and anti-protons. They slowed the particles down, allowing them to join together and form the first **anti-atom**. They produced the

first ever particle of **anti-hydrogen**! Physicists are still trying to find out if anti-matter elements have exactly the same mass as matter elements, or if the polarity of the electrical charge is the only difference. It is expensive too. The cost of producing one gram of anti-hydrogen is estimated to be **trillions** of dollars!

When the universe was formed, it is believed that roughly equal amounts of matter and anti-matter were created. Almost all the matter we see or detect now is 'normal' matter, which is not easy to explain.

So where has all the anti-matter gone? This is a serious question for physicists and one of the great problems in our understanding of the universe.

Gamma rays and health

Gamma rays are used in **radiography** departments in hospitals, both diagnostically and therapeutically.

'Diagnostic' means to find out what the problem is. Diagnostic radiography uses different kinds of radiation to 'see' inside the human body. We have already discussed x-rays.

Gamma rays possess much more energy than x-rays which means that they can pass through skin, fat, muscle and bone. This makes gamma ray sources more difficult to contain and store. We need to use a very dense material to stop gamma rays and so containers are normally made of lead, which is a very dense metal.

Diagnostic and therapeutic techniques

1   One common diagnostic technique involves using a substance, which emits gamma rays, as a **tracer**.

Tracers are radioactive liquids which may be injected (by syringe), ingested (eaten or drank), or inhaled (breathed in as a gas) by the patient.

The radioactive tracer inside the body then emits gamma rays, most of which pass through and out of the body. However, different kinds of body tissue, such as skin, muscle, fat and bone will absorb different amounts of the gamma radiation. A special gamma camera, linked to a computer, will detect the different amounts of gamma radiation emitted as the tracer travels round the body via the bloodstream or via the digestive system. Accurate images of organs or blood flow may then be created.

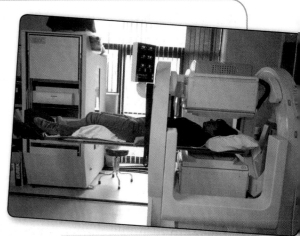

A patient in gamma camera

The tracers used do not stay radioactive for long. This ensures that the radioactive substance isn't present in the body for too long and doesn't harm healthy cells or body tissue.

2 More recently, **PET** (**Positron Emission Tomography**) has been used in some hospitals.

PET scans show even more detail than x-rays and CT scans.

A radioactive source which produces positrons (anti-matter versions of electrons) is injected into the patient. This tracer accumulates in the desired tissue, where it emits positrons.

Remember that when matter meets anti-matter, both particles are annihilated and gamma rays are emitted.

These are then detected by the gamma camera, and so an image can be built up.

3 **Therapeutic radiography** uses the property that some radiations can damage and kill human cells. Doctors use this technique to kill or shrink many cancerous tumours.

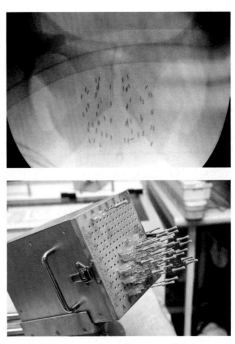

4 **Teletherapy** uses a device which fires a beam of gamma rays into the body from a distance and aims them at cancerous cells. As the rays pass through the body, they **ionise** some of the atoms in the cancerous cell, damaging or even killing the cell. Care is taken to rotate the beam around the patient, whilst always aiming the beam through the tumour. This means that the tumour receives a large dose of radiation, whilst the surrounding healthy tissue receives only a small dose of radiation.

5 **Brachytherapy** involves placing a radioactive source inside the body, close to unhealthy tissue.

It is usually injected in liquid form, but sometimes a piece of radioactive material is inserted into the body, near the tumour where the radiation then kills the cells. It is then removed.

X-rays

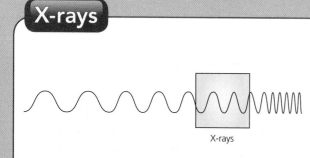

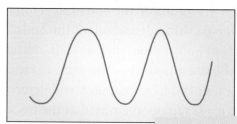

X-rays lie between gamma radiation and ultraviolet radiation on the Electromagnetic Spectrum

The region of the Electromagnetic Spectrum referred to as x-rays lies between ultraviolet radiation and gamma rays.

The wave with the 'X' factor

Use your fingers to cover a lit torch.

It looks red because some light is passing through your skin.

Imagine if you could turn up a special dial on the torch to give the light more energy – so much energy in fact, it could shine right through your skin and muscles, leaving the shadows of the bones inside your hand.

We **can** make waves like this. They are invisible, high frequency, high energy electromagnetic waves called **x-rays**.

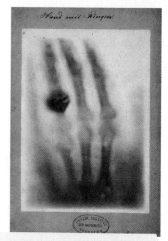

Wilhelm Röntgen: X-man

In 1895 a German scientist, Wilhelm Röntgen, wrote about discovering *a new kind of light* which he called x-rays (x for unknown). He found that the rays would pass through cardboard, books and even his wife's hand! The x-rays passed through her skin and muscle and carried on to reach some photographic paper. This caused the photographic paper to darken in most areas. The x-rays could not pass through her bones, however, and a light skeleton 'shadow' of Mrs Röntgen's hand could be seen on the paper!

The scientific name for an x-ray photograph is a **radiograph** and radiography is an important part of modern healthcare.

Xtremely useful!

X-rays may be used to diagnose some problems inside the human body without needing to cut a patient open.

An x-ray detector is placed behind the patient and x-rays are then directed at the patient.

- Where they strike bone, the x-rays are absorbed and don't go any further.

Wilhelm Röntgen's first radiograph (x-ray) of his wife's hand, from 1895

$\Rightarrow$

X-rays

- Where they strike skin, muscle and organs, the x-rays pass through and travel on to the detector. (Nowadays an electronic detector is used and the images obtained are shown on a computer screen, not photographic paper.)

Using the detector, we can build up a picture of the bones, or other dense foreign objects, inside the body.

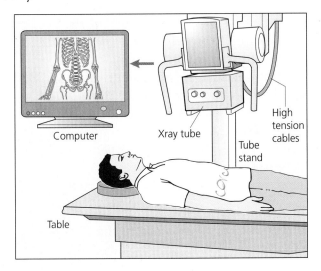

Computer   Xray tube   High tension cables   Tube stand

Table

As well as being used to look at bones, these rays may be used in other ways too. For instance:

- In hospitals, **CT scanners** build up x-ray images into 3D pictures.

- Dentists may use x-rays to look at teeth.

- Engineers may use x-rays to find cracks in materials such as aeroplane wings and gas pipes.

- Astronomers may look at x-rays given out by black holes and quasars.

Xercise caution!

It used to be common for people, including young children, to have x-rays taken of their feet when they went in to a shoe-shop to buy new shoes! In the 1960s an x-ray machine was kept in the shop. People didn't realise the risks involved at the time.

X-rays can be very dangerous, and special care must be taken when using them.

They may damage cells (sometimes causing **mutations** to form cancer) or even kill cells. Great care is taken now to ensure that patients who require an x-ray are kept safe by using protective lead aprons.

Lead is a dense metal which does not allow x-rays to pass through it and so the aprons can prevent harmful radiation from reaching parts of the body which do not require to be examined.

This patient is wearing a lead apron to protect them from the x-rays

However, this dangerous property of x-rays may actually be useful. Beams of very high power x-rays, known as *hard x-rays* may be used to destroy cancerous tumours.

QUESTIONS

1 What are x-rays?

2 How can they be used to examine patients internally (without cutting them open)?

3 Why do you think the NHS changed to using computer or digital images rather than photographic films?

4 a) Plan a simple experiment to demonstrate how x-rays are used to produce images.

   OR

   b) Write a paragraph on the uses of soft and hard x-rays in medicine.

Active Learning ▶

Activities

1 A new way of using x-rays has been developed by scientists. A technique called *x-ray backscatter* may be used to look not through skin or muscle but just through clothing. This type of scanner could be of great use in Airports to detect hidden weapons or illegal substances. However, Human Rights activists complain that this is a breach of privacy. Would you like a security person seeing a detailed outline of your body under your clothes?

   What do you or your group think?

2 Patients normally have to wait a few weeks before they can get an appointment for a CT or MRI scan at hospital. Private companies have bought CT and MRI scanners, and they allow people to purchase scans and get an appointment within days or even hours. Some people see this as progress, but others feel that these companies are preying on a person's fear.

   Would you pay to go for a health check if you felt fine?

   Would you pay to go for a health check if it was advertised that it could detect illnesses before they became serious?

Ultraviolet radiation

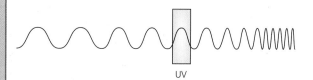

UV

Ultraviolet radiation lies between x-rays and visible light in the Electromagnetic Spectrum

The ultraviolet region of the Electromagnetic Spectrum is to be found between visible light and x-rays. Ultraviolet rays have higher frequencies than visible light but lower frequencies than x-rays. The word 'ultra' is Latin for 'beyond' and UV rays lie beyond light in the spectrum.

Background

Johann Ritter was a German physicist and chemist who made one of the earliest types of battery and developed the technique of *electroplating*. He is also the man who, in 1801, discovered the **ultraviolet** part of the Electromagnetic Spectrum.

⇨

He was experimenting with a chemical called silver chloride which turned black when exposed to light. He wondered if the colour of the light might have any noticeable effect on the reaction.

He then used a glass prism to split white light into the visible spectrum, and placed samples of silver chloride at each colour in turn. He noticed only a weak reaction near the red end of the spectrum. This steadily got stronger as he placed samples in the orange, yellow, green, blue and indigo sections of the spectrum. Violet producing the strongest reaction of all the visible radiations.

He had heard about Herschel's discovery the year before, of an invisible light outside the red end of the spectrum, called **infrared**. Ritter, also knowing that the Scottish physicist James Clerk Maxwell had predicted other kinds of light outside of the visible spectrum, wondered if there might be some similar kind of invisible light beyond the violet end of the spectrum.

He placed a sample of silver chloride just outside the violet end of the spectrum, and was amazed to see it produced the most intense reaction of all!

He called this invisible light 'chemical rays', presumably because it caused a chemical reaction in the silver chloride. This type of light later became known as **ultraviolet light.** More correctly, it is called **ultraviolet radiation**.

Types of Ultraviolet radiation

Our Sun produces the entire range of the Electromagnetic Spectrum.

Some of the ultraviolet (UV) radiation from the Sun gets through the layers of our atmosphere and reaches us on the Earth's surface.

Astronomers tend to split ultraviolet radiation into three main bands. These are called *near*, *far* and *extreme* ultraviolet.

However, you may be more familiar with the terms UVA, UVB and UVC, which are used to describe these different types of UV in sunlight which affect humans.

Stay safe in the Sun!

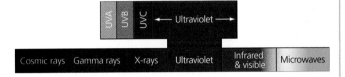

Ultraviolet radiation can be divided into UVA, UVB and UVC bands.

Of the three UV types, only UVA and UVB manage to penetrate the Earth's atmosphere.

In fact, our atmosphere stops around 98% of all UV rays from the sun. All of the UVC and large parts of UVB are absorbed by oxygen and ozone molecules in our atmosphere.

Ultraviolet light is used by our bodies to help produce **Vitamin D**, and we get all that we need from a normal, everyday exposure to sunlight.

Most people will know that UV is responsible for tanning skin, and that tanning is a risky business! A suntan might make you look like you have a healthy glow, but it is actually your body's reaction to **damage** from ultraviolet radiation.

UVA penetrates deeper into your skin than UVB. It damages collagen fibres in your skin, leading to premature aging (wrinkling) of the skin. UVA destroys Vitamin A in the skin, which may cause further damage. UVA is also known to cause allergic reactions and skin rashes.

UVB is the cause of most sunburn. UVB penetrates the top layers of your skin where it can cause genetic damage by breaking DNA bonds in your skin cells. Your skin cells can repair a lot of this damage using proteins in the cell nucleus, but if you over expose yourself to UV light, too much damage can be done. The broken DNA bonds can lead to the formation of cancers such as melanoma, squamous cell skin cancer and basal cell cancer.

Skin cancer statistics

It is clear that many people still want to be tanned, even after Health warnings. According to the HPA (Health Protection Agency), the number of people developing skin cancer is rising each year in the UK, with around 40 000 new cases and around 2 000 deaths every year!

Around 90% of these are linked to UVB exposure.

Teenagers, young children and especially babies are at particular risk. They need to have their skin protected when they are young. This may reduce the risk of developing cancer and eye-disease later in life. Severe sunburn in children has been linked to melanoma (the deadliest form of skin cancer) according to the United States Food and Drug Administration.

In conclusion it is worth pointing out that NASA state on their website that:

'Even careful tanning kills skin cells, damages DNA and causes permanent changes in skin connective tissue which leads to wrinkle formation in later life.

There is no such thing as a safe tan.'

It's important that you know how to stay safe in the sun!

Remember that your body tries to protect itself from sun damage by providing a pigment in the skin called *melanin*. This can absorb the UV and dissipate it as heat. As it works, melanin darkens the skin. This is your suntan! The 'healthy glow' you may desire is actually your body's cry for help!

Eye damage

Sunglasses that protect against UV radiation should also be worn. Your eyes are also susceptible to UV damage.

The cornea in your eye is particularly good at absorbing UV, and because there is no blood flow to it, it is largely unable to dissipate heat. Skiers have to wear goggles because they can suffer temporary bouts of 'snow blindness' where excess amounts of UV reflected from the snow and ice cause their corneas to cloud over temporarily.

Overexposure of the eyes to UV has also been linked to the permanent clouding of the cornea, called a **cataract**.

Ozone

We should be thankful that the *ozone layer* is protecting us from almost all of the UV radiation flooding towards us from the Sun (and other objects in space). However, the use of chemicals called CFCs (chlorofluorocarbons) has created a hole in the ozone layer. These chemicals destroy ozone. Fortunately the use of CFCs has been banned in an effort to save our UV shield.

'Nature red (well….ultraviolet) in tooth and claw'

Many insects, for instance bees, 'see' using ultraviolet radiation. The world would look a whole lot different through a bee's eyes (and not just because they are compound, multi-lens organs).

If **we** look at a flower we might for example see it as yellow.

If we were to look at the same flower using ultraviolet light, as a bee would see it, the yellow would be gone, and instead we might see bright bluish white markings. These are like 'road signs' telling bees where the pollen is.

Some breeds of scorpion have evolved pigments in their chitinous armour which reflect UV light. Bees are attracted down to the bright ultraviolet lit object, and the scorpion has a tasty, nutritious meal delivered straight to him without having to get off the sofa.

The ultraviolet catastrophe

Towards the end of the nineteenth century many students were being dissuaded from studying physics!

It was assumed then that our knowledge of Physics was complete and there was nothing else to learn! Lord Kelvin infamously said:

'There is nothing new to be discovered in physics now, all that remains is more and more precise measurement.'

However, there *was* just one more thing to be explained. That was the colour of hot objects.

- When we heat up an iron nail, it glows red.

- When we heat it up more it turns yellow.

- When we heat it up even more still it glows white hot.

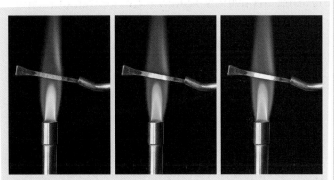

As the nail gets hotter, it glows red, then yellow and then white

Classical physics (as it is now called) just could not explain it. It predicted that the amount of energy would be infinite at the high-energy end of the spectrum. As x-rays and gamma rays had not yet been discovered, the highest energy EM wave known was ultraviolet radiation. This breakdown in physics became known as the *ultraviolet catastrophe*.

Fortunately, it took only a few years for Max Planck and Albert Einstein to help explain it using *modern physics*, or as we call it now, *quantum physics*.

The colours of hot objects are made up of different frequencies of light mixed together.

If you have any questions at all about quantum physics, just ask your physics teachers – they will be glad to explain it all to you!

Some applications

Health

1   Phototherapy is the use of controlled exposure to UVB to treat skin conditions such as psoriasis and eczema. Great care is taken to supply the dose that will treat the condition but not overexpose the patient and cause sunburn.

2   Dentists use UV light to 'cure' or harden newer types of dental fillings. The process, called photopolymerisation, uses UV light to cause strong chemical bonds to form in a white polymer paste, which hardens, filling the tooth.

3   Ultraviolet radiation is used to sterilise equipment in hospitals. Materials are placed in a tank and are given a large dose of UV. Just as UV light can damage cells in our bodies; it can do the same to germs, killing them and sterilising the equipment. The same process is also used to treat food and drinking water.

Fly killer

Some shops, especially those selling food, have ultraviolet flykillers. These devices use UV lights to attract any insects in the shop away from the food. High voltage wires are placed near the UV lamp which electrocute and kill the fly when it touches them.

Security

Bank notes have special dyes within them which only show up under UV rays. This helps to tell whether the notes are genuine or forged. UV security pens work in a similar way. They use a dye which is invisible under normal lighting, but which shows up under a UV lamp. You can write on the object without it ruining the look of the object. The police advise you to mark objects like bikes and mobile phones with your house number and postcode to help identify them if they are stolen or lost.

Ultraviolet and astronomy

Astronomical observatories which make use of UV radiation need to be in space because our atmosphere stops around 98% of UV radiation from reaching the Earth's surface. Ultraviolet light, and the ability to detect it, is of huge interest to astronomers. It provides yet another way of seeing the hidden universe.

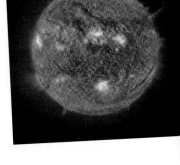

UV is of particular interest to those interested in the *evolution* of the universe. UV light is given off by the *hottest* and *most active* objects, like young stars. Older stars tend to give off red and yellow light which means they are cooler (still millions of degrees Celcius, however!).

$\Rightarrow$

Ultraviolet radiation

Have a look at the following image:

The UV image shows the younger (hotter) stars and the visible light image shows the older (cooler) stars. By comparing the images, astronomers can find out a lot about how galaxies form and evolve.

GLOSSARY

Anti-matter a strange type of matter that is not easily found

Black hole the remains of a star where no light can escape

CT scanner a medical device used to image parts of the body

Gamma rays high frequency EM radiation

Infrared EM radiation just beyond the red end of the spectrum

Ionise to gain or lose a charged particle

Mutation a change that creates a new genetic character or trait

Radiograph the scientific name for an x-ray photograph

Radiography the use of x-rays to view the inside of the human body

Supernova an explosion of giant stars

Tracer a radioactive chemical which is used to assist diagnosis of patients

UV rays (ultraviolet) EM radiation which can cause sunburn

X-rays high energy EM radiation which can detect internal problems

FORCES, ELECTRICITY AND WAVES

Waves

3

Radio waves

Level 2 — What came before?

● SCN 2–11b

By exploring reflections, the formation of
shadows and the mixing of coloured lights,
I can use my knowledge of the properties of
light to show how it can be used in a creative
way.

Level 3 — What is this chapter about?

● SCN 3–11b

By exploring radiations beyond the visible,
I can describe a selected application,
discussing the advantages and limitations.

Radio waves

In the last chapter we looked at the high frequency end of the Electromagnetic (EM) Spectrum. In this chapter we look at the other end of the spectrum where we find low frequency, large wavelength, radio waves.

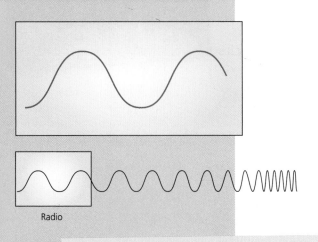

Radio

Radio waves have the lowest frequency and the longest wavelength

Radio waves are used for communication from long range radio stations to short range police radios. As this chapter shows, there are other uses too.

The first person to send and receive electromagnetic waves was the German physicist Heinrich Hertz. He didn't seem excited by what he had done and although he said that this proved that James Clerk Maxwell was correct, he went on to say that,

'It's of no use whatsoever…we just have these mysterious electromagnetic waves that we cannot see with the naked eye.'

The Italian, Guiglielmo Marconi was just a teenager when he read about Hertz's successful experiment, and he and others could see immediately a use for such waves. Both Marconi and the Croation scientist Nikola Tesla pioneered work on radio waves as a means of communication.

Initially Marconi got the patent, but in 1943 the US Supreme Court overturned the decision and stated that Tesla was the inventor of radio. Unfortunately, Tesla had died penniless earlier that same year.

Guiglielmo Marconi

Scientific summary of radio waves

- *Transmission:* Radio waves are produced (or transmitted) by making charged particles move in an aerial.

- *Detection:* Radio waves are detected because they make charged particles in another aerial move. This movement produces a small electric current, which is then detected.

We use aerials therefore to transmit (send out) and receive (pick up) radio waves.

In the EM Spectrum, 'radio waves' is the term given to waves that are used for broadcasting by radio stations like BBC Radio Scotland, Radio 1, Clyde 1 FM and West FM. Radio waves are also used by:

- The Police and other emergency services, who use a certain range (or band) of frequencies.

- Citizens Band (CB) and amateur 'ham' radio enthusiasts, who use other bands of frequencies in the spectrum.

- Television broadcasters, who use high frequency radio waves to transmit TV signals – sometimes we refer to these as TV waves.

We are most familiar with radio waves being used for long distance communication (telecommunication).

The following table shows different radio frequencies and their properties.

A radio station can be identified by its wavelength or its frequency.

For example, BBC Radio Scotland broadcasts on a range of frequencies, from 92.4–94.7 MHz (often announced as 92–95 FM). These frequencies are in the VHF band.

Radio wave band	Frequency band	Wavelengths (Wave band)	Use	Range	Means of travel
Long Wave	30 kHz–300 kHz	10 km–1 km	LW Radio	10 000 km	Diffraction
Medium Wave	300 kHz–3 MHz	1 km–100 m	MW Radio	1 000 km	Diffraction
Short Wave	3 MHz–30 MHz	100 m–10 m	Shipping, CB	10 000 km	Reflection by the ionosphere
VHF	30 MHz–300 MHz	10 m–1 m	FM Radio, Police	10 km	Straight Line
UHF	300 MHz–3 GHz	1 m–0.1 m	TV	100 km	Straight Line

How radio waves travel

Radio waves can be sent over long distances using different methods. As shown in the last table, these are **diffraction**, **reflection** by the ionosphere and by **transmission** in a straight line.

1 **Diffraction** is the bending of waves around (or over) an obstruction.

This means that the transmitter does not need to see the receiver for the waves to be picked up.

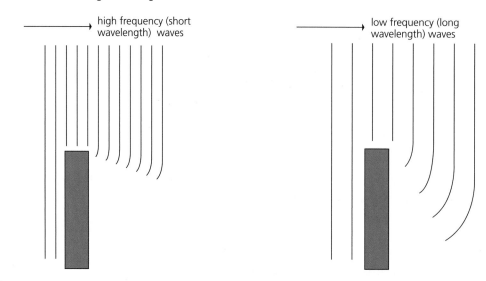

high frequency (short wavelength) waves

low frequency (long wavelength) waves

How radio waves travel

High frequency waves diffract (bend) less than low frequency waves.

Or in other words, short wavelength waves diffract (bend) less than long wavelength waves.

This explains why in some places, especially in the Highlands, people can get radio reception, but no TV reception.

Radio waves have lower frequency than TV waves, and this allows radio waves to diffract (bend) easily around buildings and over hills to reach a radio receiver.

However, TV waves have higher frequency than radio waves, and they cannot diffract as much. This means that TV stations need to have the transmitters high on hilltops so they can reach as many TV aerials as possible.

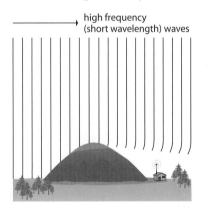 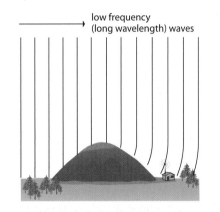

2 **Reflection by the ionosphere** is caused by radio waves reflecting off charged particles (electrons) high up in the atmosphere. This method is not always ideal because fluctuations in the ionosphere and weather conditions may affect the distance a signal can travel.

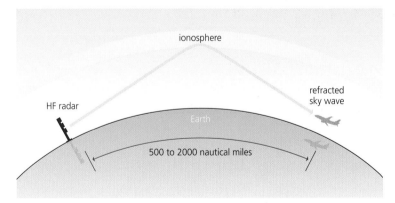

3 **Transmission in a straight line** means the transmitter must 'see' the receiver or there are boosters and transmitters along the hill tops. This happens with mobile phone masts.

Another way is to send the radio wave (or TV) signal to a **satellite** above the Earth where it is boosted and sent back to a different place on the Earth's surface.

How a radio works

Sound waves such as your voice or music can be heard, but cannot travel large distances. Even though sound travels fast compared to say cars and trains, it is as slow as a snail, compared to the speed of light!

Radio waves on the other hand can travel large distances, and like all EM waves, they travel at the speed of light. However, the problem is that we can't hear radio waves.

If only there was some way to combine the best of both waves?

Fortunately, there is!

In a process called **modulation** radio stations take the sound waves from the music or DJ and combine these audio waves with a radio wave. They broadcast these coded radio waves which can travel huge distances at the speed of light. The modulated waves are then decoded by your radio so you can hear the original sound wave again.

Any radio can be broken down into six main parts. These are:

- aerial
- tuner
- decoder
- amplifier
- electrical supply (to power the amplifier)
- loudspeaker.

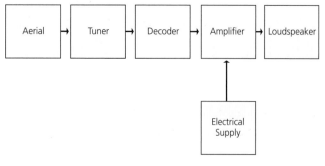

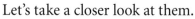

Let's take a closer look at them.

1 Aerial

We are constantly bathed in radio waves broadcast by radio stations. They are invisible to us, but they are all around us, and we absorb them.

Likewise, a metal aerial absorbs electromagnet waves reaching it.

However, aerials pick up (absorb) many EM waves, not just one from a particular station!

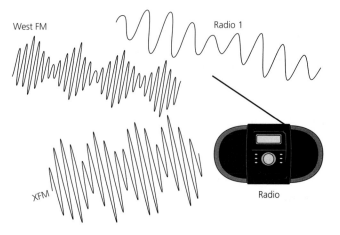

Electrical sparks emit strong electromagnetic waves.

If we de-tune a radio and create some electrical sparks, the EM waves produced are picked up by the aerial on the radio and heard as a crackling sound.

2 Tuner

The tuner selects **one** radio station from the many that are picked up by the aerial.

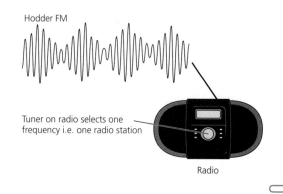

3 Decoder

When we **decode** a signal we basically chop it in half!

We want to lose the bottom portion of the signal. This allows the original sound wave to be used.

The following graphs show what would be seen on an oscilloscope screen before and after decoding. (The oscilloscope shows voltage on the *y*-axis and time on the *x*-axis.

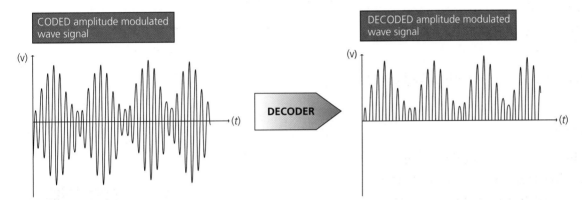

Try tracing a line over the top of the peaks of the decoded wave, to see the original sound wave appear.

4 Amplifier

The amplifier takes the weak decoded signal and gives it energy (from the electrical supply). This boosts the signal, giving it more energy.

In the following diagram, we can see the effect on the wave as it receives energy in amplification.

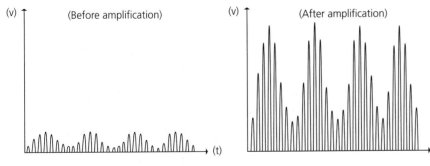

You will see that only the amplitude of the wave increases. Its **frequency** stays the same.

Imagine that the amplifier increased both the volume **and** the frequency. Now imagine you are listening to your favourite song, but it's just too quiet.

If both the amplitude and the frequency increased when you turned up the amplifier, what would happen to the sound?

It might be funny for a wee while, but pretty soon it would get annoying…

5 Electricity supply

If we want to make the decoded music or speech louder, we need to give the electrical signal more energy.

We can't just pull the energy out of thin air. We need to use a power supply.

The electricity supply to a radio is either obtained from a battery or by plugging it into the mains.

6 Loudspeaker

The loudspeaker takes the amplified signal, and changes the electric energy into sound energy.

Other uses of radio waves

1 Radio astronomy

Radio astronomy is a well-established way of using radio waves to 'see' into space. Many objects in space emit energy not as visible light, but as radio waves! These objects would be invisible to us without radio telescopes like the Lovell Telescope in Jodrell Bank.

The image above shows the same galaxy captured by an x-ray observatory, a radio telescope and an optical telescope. The three images are combined to create a composite image.

Radio telescopes are more useful than optical telescopes for trying to find black holes and quasars.

The radio telescope shown has an enormous 'dish'. The radio waves arriving from space are very weak. They have to be collected over a large area and brought to a focus, which strengthens them. An aerial at this focus collects them and converts them into electrical signals.

The Lovell Radio Telescope dish at Jodrell Bank observatory, in Cheshire

The larger the collecting area the more sensitive the telescope will be and the weaker the waves that can be received. The Lovell Radio Telescope has a dish which is 76m in diameter. It is in fact a concave reflector.

The radio telescope shown below is the largest in the world. It is the Arecibo Radio Telescope in Puerto Rico. It is built into the crater of an extinct volcano and the dish is 305m across.

China is currently building what will be the world's largest radio telescope. It should be ready in 2013 and will have a dish roughly the size of 30 football pitches!

The Arecibo Radio Telescope in Puerto Rico

As you know, shiny objects act like mirrors for visible light and infra red.

However, a metal mesh will act as a mirror for microwaves!

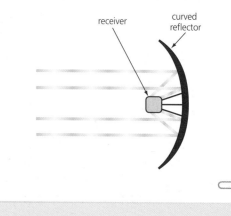

receiver   curved reflector

A satellite dish is curved like a concave mirror, and we call it a parabolic reflector (a parabola is a curve). It uses metal mesh for reflection.

Curved reflectors are used because they reflect lots of weak signals to a focus, which makes a strong signal. The antenna/aerial/microphone is then placed at the focus to detect this strong signal.

(A microwave oven uses a metal mesh in the door to reflect microwaves and stop them leaving the oven through the glass.)

2 MRI

MRI (Magnetic Resonance Imaging) uses radio waves and a magnetic field to create an image of the inside of the human body.

The patient is placed into a large and powerful magnetic field. The magnet is so strong that no metal objects must come near it. Metal being attracted could injure the patient and damage the machinery. The MRI machine has to have its own dedicated room with no metal. This poses some problems and patients must be checked to make sure they have no metal objects or metal fragments due to accidents, pacemakers or other medical procedures, inside their bodies.

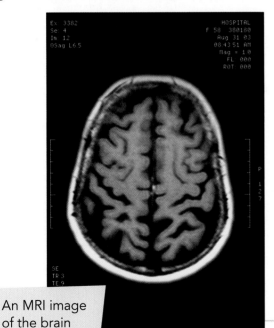

An MRI image of the brain

Our bodies contain a lot of water molecules, and water molecules contain hydrogen atoms. The magnetic field is so strong that it causes the hydrogen atoms in our bodies to become aligned in one direction.

Once aligned, the machine emits radio waves which cause the hydrogen atoms to change direction. When the radio wave are switched off, the hydrogen atoms re-align with the magnetic field. To move back into alignment with the magnet, the atoms need to get rid of energy. As they do so, the machine detects this energy.

Different types of tissue (bone, muscle, fat) contain different amounts of water. This means they contain different numbers of hydrogen atoms.

- Bone, for instance does not contain very many hydrogen atoms and so gives off very little energy – it shows up as a dark area.

- Fat, on the other hand, contains a high concentration of water (and therefore hydrogen atoms), so it shows up as a bright area. (If the radiographers say that you are very bright, they might not be complementing you.)

The machine detects the amount of energy given off by the different types of tissue and their locations, and uses this data to build up a highly accurate 3D image of the inside of a patient.

MRI produces a more accurate image of the inside of the patient's body than say x-rays or CT scanning. It is also safe, although there is a small chance it may pose a risk to developing babies, so women who are 12 weeks pregnant or earlier are not given MRI scans. There may be no risk, but there is not enough information at present to be sure.

35

GLOSSARY

Decode to separate the sound wave from the transmitted wave

Diffraction the bending of waves around or over an obstruction

Ionosphere a region of the upper atmosphere

Modulation combining sound and radio waves for transmission

Reflection to rebound from a barrier

Satellite an object that orbits a planet

Transmission the sending of waves into the atmosphere

FORCES, ELECTRICITY AND WAVES

Waves

4

Infrared radiation and heat

Level 2   What came before?

SCN 2–11b

By exploring reflections, the formation of shadows and the mixing of coloured lights, I can use my knowledge of the properties of light to show how it can be used in a creative way.

Level 3   What is this chapter about?

SCN 3–11b

By exploring radiations beyond the visible, I can describe a selected application, discussing the advantages and limitations.

Infrared radiation and heat

In this chapter we continue our journey through the low frequency part of the Electromagnetic Spectrum. Here we look at infrared radiation (IR) emitted by heated objects.

The infrared region of the Electromagnetic Spectrum lies between microwaves and visible light.

It is just beyond the red end of the visible spectrum. 'Infra' means 'below'.

William Herschel was the astronomer who discovered the planet Uranus. In 1800 he was performing an experiment with light. He had used a glass prism to split white light into the different colours of the visible spectrum, and was using a blackened thermometer to find if the different colours had different temperatures.

He was shocked however to find that the highest temperature was found just *beyond* the red end of the visible spectrum. At first he wasn't sure of what this 'invisible hot stuff' was. He had just discovered infrared waves!

Infrared waves have higher frequencies than microwaves, and as such they have shorter wavelengths. Infrared radiation shares the same properties of reflection and refraction as visible light.

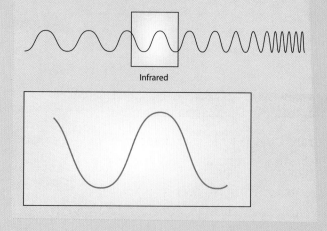

Infrared radiation lies between visible light and microwaves on the Electromagnetic Spectrum

Infrared

Infrared is commonly thought of as heat because hot objects such as charcoal in a barbeque emit infrared electromagnetic waves.

Once the flames have died down, the coals glow. They are giving off both heat and light.

After a short time, they lose energy and stop giving out visible light (they stop glowing). However, the coals will still be very hot, and as such will be producing a lot of infrared radiation.

Infrared detectors

We can use **thermographic film** to detect IR. Pressing your hand against it will cause the film to change colour.

We can also use electronic devices called *photodiodes* which detect IR radiation.

You might even have an infrared detector in your pocket!

Digital cameras and mobile phone cameras use tiny electronic chips called CCDs (charge-coupled devices). These devices are used in cameras to detect visible light, but many of them also pick up infrared waves.

Remote controls used for TVs and DVD players use infrared too. If you press a button on the TV control whilst looking at it through your digital camera or mobile phone screen, you should be able to see the IR light which is normally invisible to us. It is a simple experiment and worth trying! You should see a small dot appear on the end of the remote control. Some cameras show blue dots but it is just the way the detector operates.

$\Rightarrow$

The infrared range of frequencies is usually split into three bands – far, mid and near infrared. (Near infrared is the one closest to visible light.)

Infrared radiation can be divided into far, mid and near bands

Infrared waves which are detected as heat are from the far infrared band.

Remote controls use near infrared which isn't detected as heat by our skin. The highest frequencies of IR waves are in this band and this means they don't diffract well. Hence the reason why you might need to wave the TV remote about a bit to get it to work if something is in the way!

Plants

If you look at green leaves on a plant using an infrared camera, you will see that they appear to glow.

Plants need UV light for **photosynthesis** to happen, but IR light is of no use to them. In fact it could cause the plant to overheat if it was absorbed. Chlorophyll in the leaves reflect infrared light away from the plant.

Animals

Snakes of the *pit viper* family have a small pit near each eye. Their eyes allow them to detect visible light just like humans. However, they have evolved pits near the eyes which allow them to detect infrared light too. This means they can 'see' warm objects in the dark. This gives a big advantage when they hunt warm-blooded prey like mice and rabbits in dark burrows!

Applications of infrared radiation

Health

Infrared light is used in various ways in hospitals.

1 Cancerous tumours tend to be warmer than other body tissue. This is because the tumour diverts blood flow to itself to help its growth. Thermal imaging cameras can be used to scan a patient's body, invisible IR light is detected, and an image can be built up. This can be used to detect a tumour in a patient's body.

2 Physiotherapists use IR treatment to help repair torn muscle tissue. Muscles repair themselves more quickly when they are heated. This is one reason a nice warm bath is so relaxing.

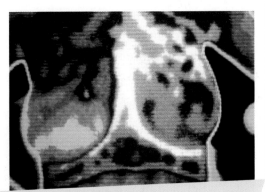

An infrared image of a tumour showing increased skin temperature over a cancerous left breast. The blue areas are the coldest and the yellow areas are the hottest.

Thermal imaging cameras

These special cameras which detect heat are used by emergency services. Police use them to help find criminals who are hidden or in the dark. Fire fighters use them to assess fires and to find trapped victims. Specialist teams also go to sites of earthquakes, tsunamis and mudslides around the world to help find people trapped by debris.

The cameras respond to living human bodies giving off heat, and they are sensitive enough to detect the IR waves even through metres of rubble!

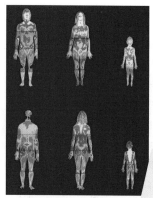

An infrared image of the human body

Satellites

Satellites fitted with IR detectors are used to monitor the weather. These satellites are in **geostationary orbits**, which means that even as they as they orbit the Earth, they always stay above the same part of the Earth. Weather satellites monitor land, sea and clouds by infrared, allowing them to 'see' the different temperatures of each. Warmer objects give off more infrared radiation. The images are then converted to give an image like the one below.

An infrared image of Scotland

The lighter parts are colder, meaning these are cold fronts and rainclouds.

Another use is for satellites to image vegetation using ir. These satellites scan crops using a technique called *IR spectroscopy*. It can show up diseases which would not be obvious to the farmer down on Earth.

Astronomy

Infrared astronomy like radio astronomy can be used to show objects not visible to the naked eye.

Recently, the European Space Agency (ESA) launched the Herschel Space Observatory satellite. This is the newest addition to infrared astronomy. It was going to be called the *Far Infrared and sub-millimetre Telescope* (FIRST for short), but it was decided to honour the man who discovered the infrared part of the Electromagnetic Spectrum.

The Herschel Space Observatory will study the formation and early evolution of **galaxies** in the early universe, examine the formation of stars and observe the chemical composition of the surfaces and atmospheres of planets, moons and comets.

This photograph shows the same image of space in visible light and in infrared

Other infrared telescopes are already looking out into our universe and helping to build up our understanding of it by allowing us detect things which were previously invisible to us.

Communications

Near infrared is used in fibre optic communications systems as well as visible light. Fibre optic devices are dealt with later in the book.

Heaters

You may have noticed that many gas and electric heaters have curved shiny surfaces behind their heating elements. This is similar to the way that car headlamps and torches are designed, and for the same purpose.

If you place a heating element at the focus of a curved reflector, the waves can be reflected into an outgoing parallel beam. This helps to 'shine' a beam of heat into the room, just like a beam of light from a torch.

Can cook, will cook!

How does heat energy pass from one object to another? Well, it's as easy as 1, 2, 3!

1 **Conduction** is the transfer of heat energy from one part of a substance to another part by direct contact between the particles (atoms or molecules) within the substance. This method of transferring heat energy can also happen when two or more substances make contact. Conduction works best in **solids** because the particles are closely packed together allowing them to vibrate against each other, transferring heat energy. Conduction in liquids is less effective because the particles are slightly further apart and they can move about. In gases, conduction is very difficult because the particles are so far apart they rarely collide with each other. So the possibility of transferring energy by contact is greatly reduced.

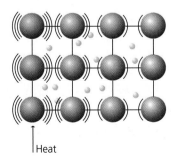

Heat

2 **Convection** is the process that transfers heat energy in **fluids** (liquids and gases). When part of a fluid is heated it becomes less dense than the surrounding, cooler fluid. Less dense fluids, like warm air, move upwards as they are more buoyant in the surrounding, cooler fluid. This causes cool, denser fluid to flow in to fill the space left by the hotter fluid. It too, when heated, will move upwards.

If there is a constant source of heat then this process will continue and a continuous movement of fluid – a **convection current** – will result.

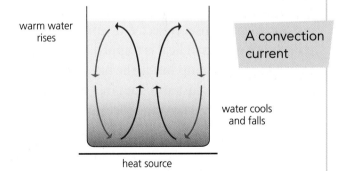

warm water rises

A convection current

water cools and falls

heat source

Both conduction and convection require particles to transfer heat energy. Heat can be transferred however even when no particles are present, by a third method called *radiation*.

3 **Radiation** is the process where invisible wave energy is given out (emitted) in all directions by warm or hot objects. These invisible waves are called infrared waves (IR waves). They are that part of the EM spectrum discussed in Chapter 1. Infrared waves can travel through empty space and so can carry energy from one place to another without the need for particles to be present. When IR waves are absorbed by another object, the wave energy changes into heat energy.

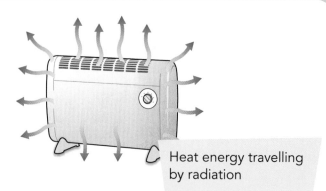

Heat energy travelling by radiation

A **solar cooker** is a device which makes effective use of all three methods of heat transfer. Used in developing countries, it offers a sustainable way of cooking food and purifying water where domestic electricity and gas are unavailable.

The Sun's rays are reflected by the curved mirror onto the dull dark cooking pot placed at the focus, where the rays are concentrated and available heat energy is greatest.

The dark pot absorbs infrared radiation from the Sun.

After absorbing the radiation, the pot **conducts** heat to the food inside which then cooks using **conduction** and **convection**.

The clear plastic bag makes positive use of the *Greenhouse Effect* – no global warming here! Infrared radiation passes into the bag and the pot but any radiation reflected or emitted by the pot cannot escape so heat is trapped in the bag. Heat energy builds up inside the bag and the temperature rises, just as it does in a greenhouse, further helping the cooking process.

A tight fitting pot lid can reduce heat loss by convection.

Solar cookers work best in sunny locations and for maximum performance should be tilted perpendicular to the Sun's rays.

The meaning of heat

Heat or hotness?

In our everyday lives we use the words 'heat' and 'hot' to describe lots of different things. For example 'Is this heat going to last?' is something you have probably heard on a warm summer's day . . . though probably not that often in Scotland!

However, scientists have a very specific use for the word heat. To scientists, heat is a type of **energy** and temperature is measure of how hot or cold something is!

$\Longrightarrow$

It's all relative

So how do we know how hot or cold something is? Can we tell by touch?

Take two cans of fizzy juice which have been out of the fridge for a while and two bowls of water: one filled from the cold tap, the other from the warm tap. Put one hand in the cold water and the other in the warm water for one minute then grab a can in each hand.

Do you think then that we can tell accurately how hot or cold something is by touch?

We use a thermometer to measure how hot or cold an object is. Thermometers come in many shapes and sizes but most importantly they don't rely on our senses to measure temperature.

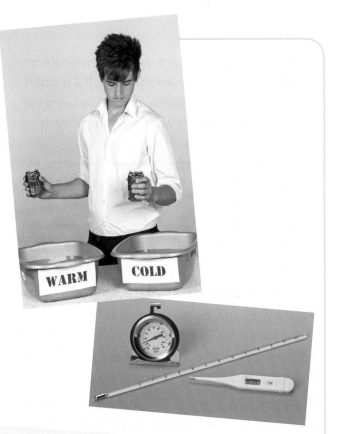

Heat and temperature

Which has more heat energy, a swimming pool of water at 25 °C or a kettle of water at 100 °C?

You've been using the same water heater; it's not a trick!

The larger volume of water takes longer to get hot because you have to supply more heat energy to it. Imagine how many kettles you would have to boil to get a swimming pool of cool water even a little hotter! Obviously you would require many more than the one kettle of boiling water in the question above!

To help you answer this question you can carry out your own investigation. All you need is a water heater from the Physics lab, a small beaker of water and a large trough of water. Carry out a fair test to increase the temperature of the water in both containers. Which one gets hotter faster?

The swimming pool contains much more heat energy than the kettle, even though it's not as hot. That is because heat is a type of **energy** but temperature is a measure of how hot something is. They are related to each other, but they are not the same thing!

Lord Kelvin

Temperature has many different scales of measurement, each with its own reference points. You are probably familiar with the Celsius and Fahrenheit scales.

Lord Kelvin of Largs, real name William Thomson, created the **Kelvin scale**. This is a better temperature scale than others because it doesn't have any negative numbers to confuse people.

Lord Kelvin was one of the world's greatest and most influential scientists and his work helped us understand the behaviour of the tiny particles that make up everything!

Kelvin was born in Belfast but moved to Scotland at the age of 9 when his father, a mathematician, obtained a professorship at Glasgow University.

QUESTIONS

1   What is the name of the temperature scale that is based on the freezing point and boiling point of pure water?

2   The Kelvin scale has a point known as 'absolute zero', can you find out what this means?

    (Get your calculator now!)

3   Energy is measured in Joules (J). It takes about 4000J to increase the temperature of 1kg of water by 1 °C.

How much energy is required to boil 1kg of tap water with a starting temperature of 10 °C?

4   A kettle supplies about 2000J of heat energy to water every second. How long would it take to boil 1kg of water using this kettle? (That is enough for three cups of tea.)

5   It costs about 3p to boil a kettle of water. What things could we do in our homes to reduce the cost of making a cup of tea?

GLOSSARY

Conduction the transfer of heat energy from one part of a substance to another part

Convection a method of transferring heat energy in fluids (liquids and gases)

Galaxies groups of stars

Geostationary orbit an orbit of a satellite which remains above a particular point on the equator

Photosynthesis the process plants use to make sugar and oxygen

Thermographic film chemical film which responds to differences in temperature

PLANET EARTH
Energy sources and sustainability

5
Hot
or cold

Level 2 — What came before?

SCN 2–04a

By considering examples where energy is conserved, I can identify the energy source, how it is transferred and ways of reducing wasted energy.

Level 3 — What is this chapter about?

SCN 3–04a

I can use my knowledge of the different ways in which heat is transferred between hot and cold objects and the thermal conductivity of materials to improve energy efficiency in buildings or other systems.

Hot or cold

Thermowhat?

Heat and its properties form a major part of the study of Physics. Many Scientists believe that the nature of the universe and many of its mysteries can be explained using the four laws of **Thermodynamics**.

Thermodynamics?…thermowhat?

Thermodynamics is the study of heat and other energy forms involved in physical and chemical processes.

Why is heat so important…or even interesting?

My favourite experiment

Go to the freezer and grab some ice cream… it's ok, it's all in the name of scientific discovery!

No expensive equipment is needed here, just start munching!

When the ice cream hits your mouth it starts to melt, why?

Your mouth starts to cool down, why?

What would happen if you were able to resist swallowing the ice cream for 1 minute? Impossible perhaps, but if you could?

The ice cream becomes warmer and your mouth becomes colder. Eventually they will be almost at the same temperature. When warm objects like your mouth come into contact with cold objects like the ice cream, heat flows from the warm object to the cold object. This process continues until both objects are at the same temperature. When this occurs we say that the objects are in **thermal equilibrium** with each other.

The Dewar flask

Just over 100 years ago the Scottish Scientist James Dewar invented a device which slowed down the heat transfer that causes two objects to reach thermal equilibrium. He understood the principles of conduction, convection and radiation that we have learned about and used the laws of thermodynamics to design the original Dewar flask.

So how did he do it?

Firstly he wondered about slowing down conduction and convection. But how?

By making the container from a poor conductor and trying to keep the cold material away from the hot stuff of course!

There is no better way to do this than to use a vacuum! By removing all the air from the space between the hot and cold materials, heat transfer by conduction and convection is reduced greatly. A vacuum contains no air molecules!

Secondly, how did he reduce heat transfer by radiation?

Hmmm, a bit more difficult but if you remember reading about the solar cooker, heat radiation behaves like light in many ways, and just like light it can be reflected. So Dewar decided to make the container walls from shiny reflective materials and he used silver.

$\Rightarrow$

Thermowhat?

You might recognise this device as a 'Thermos' flask, which is a slightly modified Dewar flask but rest assured, Sir James Dewar – he was knighted in 1904 – was the man who made the 'Thermos' flask possible.

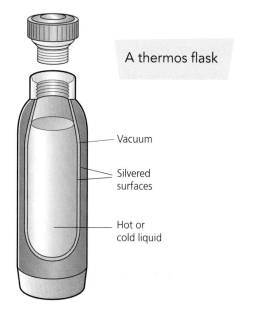

A thermos flask

Vacuum

Silvered surfaces

Hot or cold liquid

Are you the next James Dewar?

Can you explain why the temperature of a flask of hot coffee drops more quickly than the temperature of a flask of cold juice rises?

You can design a simple experiment in the lab to show this using hot and cold water, some beakers, a stop clock and a couple of thermometers.

To understand why it happens though, you need to research something called **ambient temperature**.

QUESTIONS

1 If you use a liquid-in-glass thermometer to measure the temperature of a liquid, do you really measure the temperature of the liquid accurately? Or do you measure the NEW temperature of the liquid, now that it has reached thermal equilibrium with the thermometer you have just put into contact with it?

2 What might affect the size of the temperature change produced in a liquid by adding a thermometer to the liquid?

3 Is there any way to measure accurately the temperature of a substance without changing its temperature? Start researching!

(These questions illustrate a very basic problem with measurement – not just in Physics, but in other areas too. Suppose someone is trying to measure how clever you are by giving you a written test. If you became anxious about the test, you might perform poorly, and so appear to be less clever than you really are!)

The refrigerator question!

Heat energy moves from an object with a high temperature to an object with a lower temperature

If a refrigerator door is left open, does the kitchen cool down?

Hmmm! If the refrigerator door is left open then the room must get colder. After all a refrigerator is cold and most kitchens are warm places, so the cold air from inside the refrigerator will mix with the warm air from the kitchen and this will make the room a little colder. That's essentially what the second law of thermodynamics suggests so that's got to be how it works, hasn't it?

In order to answer the question accurately we have to understand how the refrigerator cools down inside **and** what happens to the heat that we have taken away from inside the refrigerator.

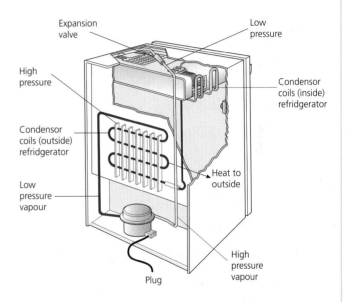

In the pipes located inside and outside of the refrigerator is a substance that alternates between a gas and a liquid known as a **refrigerant**.

- The **compressor** squeezes the refrigerant into a small space. When this happens to a gas its pressure and temperature rise.

- The refrigerant then flows through the heat exchange pipes behind the fridge and gives out some of its heat to the surroundings – usually the air in the kitchen.

- The cooler refrigerant then condenses, changing state from gas to liquid.

- The liquid is now forced through an **expansion valve** where it spreads out and occupies much more space, causing pressure and temperature to drop. It is now a really cold mist-like substance.

- The cold refrigerant then flows through the fridge and the 2nd Law of Thermodynamics works its magic. The refrigerant pipes, which are in contact with the cold refrigerant become very cold too. These pipes are in contact with the air in the fridge, and we know what that means!

The refrigerator question!

- Yes! The air in the fridge – and the food – cools down as the cold pipes and the refrigerant absorb some of the heat.

- The slightly warmer refrigerant then flows along the pipes, out of the fridge, and into the compressor where the cycle starts again!

The heat is always carried away from the food inside the fridge and released into the room surrounding the refrigerator.

(If you find the action of the refrigerator difficult to understand, consider this. When we want water to flow from low ground to higher ground, we have to use a pump. Water will not flow uphill by itself.

Neither will heat flow from a cold substance to a warmer substance unless we use some sort of 'heat pump'! A refrigerator is a heat pump.)

QUESTIONS

1 Does an open fridge really cool down a room?

2 What is so special about the refrigerant substance?

3 Why do you think the refrigerant carrying pipes outside the fridge are black in colour?

4 Using your knowledge of how hot and cold objects react when they are in contact, explain how a fridge cools down.

5 The refrigerant substances used in modern appliances differ from the substances used around 20 years ago. These were called chlorofluorocarbons (or CFCs). Can you use the World Wide Web to find out why CFCs are no longer used in refrigerators even though they work very well as refrigerants?

GLOSSARY

Ambient temperature the temperature of the substance surrounding an object

Compressor refrigerator compressors are devices that force the refrigerant fluid into a smaller space, causing an increase in pressure in the fluid and therefore an increase in the temperature of the fluid

Expansion valve a device which allows fluid to spread out, reducing the pressure of the fluid and causing it to cool down

Thermal equilibrium when objects in contact with each other are at exactly the same temperature

Thermodynamics the study of heat and other energy forms involved in physical and chemical processes

PLANET EARTH
Energy sources and sustainability

6
Energy for ever

Level 2 — **What came before?**

SCN 2–04b

Through exploring non-renewable energy sources, I can describe how they are used in Scotland today and express an informed view on the implications for their future use.

Level 3 — **What is this chapter about?**

SCN 3–04b

By investigating renewable energy sources and taking part in practical activities to harness them, I can discuss their benefits and potential problems.

Energy for ever

The wind of change

We are fast running out of fossil fuels but we need more energy than ever. Fortunately our attitude to energy production is experiencing a wind of change!

Using wind energy is the fastest growing method of producing electricity worldwide. It is free from carbon dioxide emission – the main gas associated with the greenhouse effect and global warming. Wind energy is also sustainable, reducing our reliance on fossil fuels imported from other countries and Scotland is at the forefront of this **green technology**.

Whitelee Wind Farm on Eaglesham Moor is the second largest windfarm in Europe. The 140 turbines can generate enough electricity to power 180 000 homes

Whitelee wind farm on Eaglesham Moor near Glasgow opened in 2008 and is the second biggest onshore wind farm in Europe, with 140 turbines producing enough energy to supply 180 000 homes – that's enough for the city of Glasgow! More turbines are planned for the site and the output could reach a staggering **500 Megawatts** (1 Megawatt = 1 000 000 Watts). This is equivalent to a large thermal power station using fossil fuels!

The turbines use a 3-blade system with the length of each blade around 40 metres. They can operate with wind speeds as low as 4 metres per second: however the optimum performance of the turbines is achieved with wind speeds of around 14 metres per second. Contrary to popular belief wind turbines do not work best in high winds. In fact wind speeds greater than 25 metres per second can cause damage to the turbines and so they are designed to stop working automatically at speeds in excess of this.

The following graph shows how output power is related to wind speed.

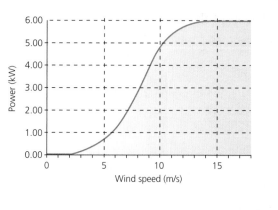

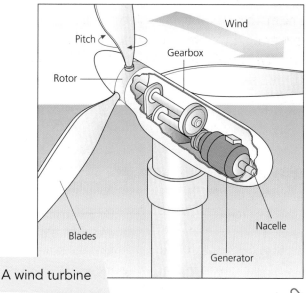

A wind turbine

The wind of change

A typical wind turbine has a set of blades rotating around a hub which is connected to the generator located in a housing called a **nacelle**. The generator, which changes *kinetic energy* into *electric energy*, is connected to the turbine blades by a gearbox. The gearing system allows the speed of the generator to be controlled effectively. When the turbine blades rotate slowly the gears enable the generator to rotate faster, which it must do to produce enough electricity.

Horizontal axis wind turbines like those at Whitelee are designed to turn fastest when they face into the oncoming wind, and the blades are set at the correct angle to ensure that the turbine gains maximum energy from the wind. This is known as 'pitching' the blades. When the wind changes direction the nacelle and blades are turned around the fixed tower to the correct position by motors controlled from a remote location.

The Whitelee project required years of planning and a crucial part of this process was a *feasibility study* which identified where each turbine should be located to make best use of the wind available at the site. Turbines must be carefully positioned so the performance of one is not adversely affected by other turbines sheltering it from the wind! The mathematical relationship between wind speed and the power it produces is called a *cubic function*. If the wind speed doubles, the turbine produces 8 times more power! So you can tell that it's all about location, location, location!

Rigorous guidelines aimed at securing the protection of local wildlife were followed during the implementation of the wind farm. Such studies are called *environmental impact assessments* and are undertaken whenever a project may have a major effect on an area. The intention is that any proposed project should have little or no impact on an area, and if it does, measures must be taken to reduce and minimise this impact.

QUESTIONS

A 3-blade wind turbine has a blade length of 40m.

a What is the diameter of the circle swept out by the blade?

b What is the area of the circle swept out by the blades?

c If you are standing in front of the turbine and a blade passes you every 4 seconds, what is the frequency of the turbine in r.p.m. (revolutions per minute)?

The tide is turning!

Scotland's unique island geography gives great hope for an emerging renewable energy. This is tidal power! The tides, a result of the earth's gravitational interaction with the moon, are a source of energy that we have only just begun to harness. Tidal power projects have received backing from the Scottish Government as we aim to reach our target of providing 50% of our electricity from renewable sources by 2020.

One of the first marine energy projects planned in Scotland is for the Pentland Firth (near the Orkney Isles) where strong and fast tides make it the perfect place to build a Tidal Power facility.

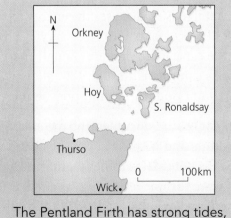

The Pentland Firth has strong tides, making it an ideal location to harness the energy of the sea

The tide is turning!

Tidal turbines are similar to wind turbines, except their rotor blades which are smaller than their wind counterparts, are located below the surface of the water to capture the energy from the tides. Unlike wind, which is unpredictable, tides are regular and predictable. The tidal turbines are therefore more efficient.

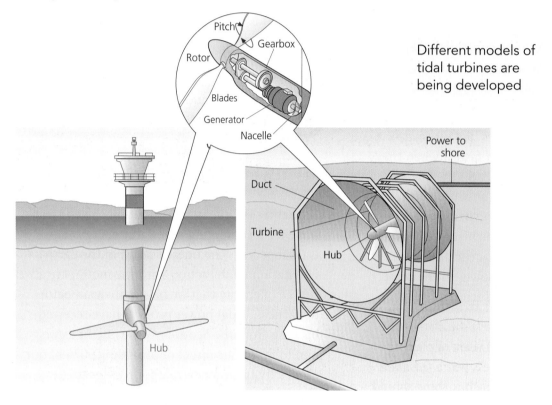

Different models of tidal turbines are being developed

Tidal turbines can be either *vertical* or *horizontal* axis devices and there are two basic energy principles behind their operation.

The vertical axis device uses the change in water levels as the tides rise and fall to convert the water's potential energy into electricity.

The horizontal axis device uses the energy from fast moving tidal currents to convert kinetic energy into electricity. Each current turbine (sorry, bad joke) has the capacity to provide enough energy for around 1500 homes and some models can rotate through 180° allowing them to work in both the ebb (outgoing) and flood (incoming) tides so increasing their generating capacity.

Marine biologists are documenting the effects of this new technology on marine wildlife and with seals, porpoises and whales common in the Pentland Firth, the effect of tidal turbines on marine habitats must not be ignored. Although wave turbine technology is a renewable energy in its early stages of development and one that is difficult to install due to the nature of working in remote areas under water, tidal power is an exciting opportunity to harness the power of the water that surrounds us!

It's our responsibility!

Whilst research into **renewable energy sources** continues we the consumer can do our bit to lower our **carbon footprint** by making our own homes as **sustainable** as possible by installing our own renewable technologies.

Solar heating panels

Many people have now installed solar heating panels. These panels are usually fixed to the roof so allowing them to absorb heat energy from the sun. This heats a special fluid in the panel that is pumped to the hot water tank where the heat energy is exchanged to the water in the tank. These systems are not **carbon neutral** because an electric pump is used to move the heat absorbing fluid around the system but they can be used to provide 30% of domestic hot water in Scottish homes, and almost all the hot water a house in Australia needs!

Wind turbines

Small-scale wind turbines are available to buy from most DIY stores but their performance is not as good as Scientists had first hoped. The good news is that Scotland is the most feasible country in Britain for this technology. However The Energy Saving Trust recently carried out a comprehensive test of small-scale building mounted turbines currently being used in Britain. It was found that on average these turbines produce 200 Kilowatt Hours of electricity per year. That's only around £25 of electricity! As each turbine costs around £1000 to install they are not really cost effective, and this is turning many consumers off.

Biomass burners

A wood burning stove not only looks great but it's an environmentally sound way of providing heat for your home. The burning wood – known as **biomass** – does produce carbon dioxide (CO_2) but this is offset by the amount of CO_2 which the tree removed from the atmosphere while it was growing. So this is perhaps a carbon neutral solution! The biomass fuel to be burned does not save money when compared with gas or oil but it is sustainable.

As consumers, we have the right to choose our own energy sources and many of us will always choose the most cost effective option, but should we not consider what's best for the environment and not just ourselves? Do we have a responsibility to create a more sustainable future? Or is what happens to us in the present all that should concern us?

GLOSSARY

Biomass organic matter suitable for use as a fuel

Carbon footprint the term used to describe the mass of carbon dioxide and other gases that contribute to global warming produced whilst carrying out a process

Carbon neutral a term used to describe a process that offsets any carbon dioxide released, for example planting trees to compensate for carbon dioxide released

Green technology a phrase used to describe examples of technology that do not impact negatively on the environment

Nacelle the housing which holds the generator in a wind turbine

Renewable energy sources sources of energy that can be used again and again

Sustainable anything that allows us to meet our needs without stopping future generations from meeting their needs

PLANET EARTH
Energy sources and sustainability

7

The nuclear power station

Level 2   What came before?

● **SCN 2–04b**

Through exploring non-renewable energy sources, I can describe how they are used in Scotland today and express an informed view on the implications for their future use.

Level 3   What is this chapter about?

● **SCN 3–04b**

By investigating renewable energy sources and taking part in practical activities to harness them, I can discuss their benefits and potential problems.

The nuclear power station

Nuclear power

At the present time in Britain, we produce around 15% of our electricity from nuclear fuel and this figure is set to increase as the British government has recently given the green light to a new wave of nuclear power stations.

The case for nuclear power

Nuclear power supporters claim that nuclear energy provides a clean way of providing our country with its energy needs as we aim to cut the amount of carbon dioxide we produce. Carbon dioxide released into the atmosphere is believed to be a major contributor to global warming. Nuclear power stations release less than 1% of the carbon dioxide produced in a fossil fuel driven power station. Another positive factor is that having much of our electric energy supplied by nuclear power stations reduces our reliance on imported fossil fuels. The growing costs and diminishing supplies of fossil fuels mean that we must look to other ways of meeting our energy needs.

Uranium

The most common raw fuel for nuclear energy production is uranium, an element found naturally in the Earth's crust. When we compare the amount of energy we can extract from similar quantities of uranium, coal and oil, we find that we can obtain two to three million times more energy from uranium! When used for electricity production in power stations, 1kg of enriched uranium produces the same amount of electricity as 14 000kg of coal!

Nuclear power stations are expensive to build and there is a lot of money spent on providing safety in case of an accident or leak, and on disposal of the highly radioactive waste. Money has to be spent also on decommissioning the plant when it closes. It is claimed however these costs are more than offset by the very cheap production cost of the electricity in a nuclear power station over a large number of years.

The statement made by Tony Blair when he was Prime Minister that 'nuclear is back on the agenda in a big way' was backed up by his successor Gordon Brown, who ordered a new system of nuclear power stations to be built in Britain. The first of these will come online by 2020.

QUESTIONS

1 What advantages does nuclear power have compared with fossil fuels?

2 Why does nuclear energy appear to be a good long-term solution to Britain's energy needs?

3 Are there any disadvantages of nuclear power?

4 What is **your** opinion of nuclear energy as a potential solution to our energy crisis?

How do nuclear power stations work?

Nuclear power stations have similar properties to coal- and gas-fired power stations. They all heat water to produce steam which is then used to turn a turbine which drives a generator producing electricity. The **big** difference is the way in which the water is heated in a nuclear station.

Well, it has changed into energy and an incredible amount of energy even with such a small mass! This energy, in the form of heat, can be used to heat water in a boiler, therefore producing the superheated steam required to turn the turbine.

The fission process also produces more neutrons and they go on to split other nuclei, so releasing more heat energy. This process continues over and over again and is known as a **chain reaction**.

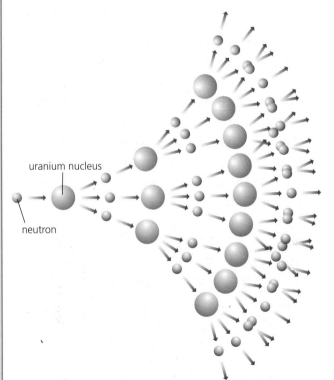

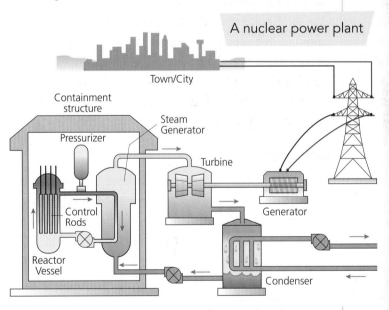

A nuclear power plant

In nuclear power stations the specially prepared or *enriched* uranium fuel undergoes a process called induced **nuclear fission**. When a uranium atom is struck by a neutron this causes the nucleus of the atom to split into two smaller parts. The amazing thing is that the combined masses of these parts is **less** than that of the original uranium nucleus! So what has happened to the rest of the mass?

The nuclear fuel is housed in a **reactor** encased in a structure often made from thick concrete and steel designed to prevent the highly dangerous radiation from escaping in the event of an accident. The gradual release of heat energy is achieved using **control rods** which keep the chain reaction under control by absorbing just the right amount of nucleus splitting neutrons!

The Chernobyl legacy

26th April 1986

At 1:23am on the 26th April 1986 the world's biggest nuclear accident occurred at the Chernobyl nuclear power plant near Kiev, now part of Ukraine but formerly part of Russia.

Whilst running an experimental test on reactor number four, a massive power surge caused the fuel rods in the reactor to rupture. This led ultimately to water in the reactor's cooling section being converted into steam. This resulted in the reaction 'speeding up' and generating more heat, incredibly quickly. The steam got hotter and hotter and with a tremendous increase in pressure. The pressure increased so much that the 2000 tonne steel and concrete roof of the reactor was blown off. This allowed radioactive particles from the reactor to escape into the atmosphere where it was carried on the wind over a huge area – including Scotland!

Casualties

The effects of the accident were immediate and approximately 1000 emergency workers attending the scene received potentially lethal doses of radiation. In the following four months, 28 people were reported to have died due to radiation poisoning but the number of people who have been affected by the disaster is much greater than this death toll suggests. The number of cases of thyroid cancer in children living in the surrounding areas rose sharply. By the year 2000 there were 4000 cases, though only nine of these proved to be fatal.

The effects of the disaster could have been reduced had the secretive Soviet Government not tried to play down the seriousness of the accident. The nearby town of Pripyat, home to the power plant workers, did not receive the order to evacuate until more than 36 hours after the accident, and a 30 km exclusion zone – still in place today – was not set up until almost one week after the accident!

The *official* estimate of eventual deaths caused by cancers directly attributable to the accident is 4000. However, an independent study by other respected Scientists in 2006 put this figure at 60 000. The exact figure will never be known.

Other problems

The impact of the disaster was felt not only by the human population. Local waterways were contaminated as were crops and grazing areas affecting livestock. Again, the true extent of the contamination and its dangers were difficult to measure but **genetic mutations** have been identified in some species, and scientists are worried these mutations will spread through the general populations of these species.

The legacy of Chernobyl remains. Many people still fear another disaster like this could happen again though the truth is that many improvements to safety in Soviet-designed reactors have been made as a result of working more openly with western Scientists. Nuclear power can be dangerous as Chernobyl has shown, but since 1986 there has not been another such serious accident at a nuclear power station anywhere in the world.

GLOSSARY

Chain reaction a repeating process where the nucleus of an atom is split, usually by a neutron, releasing energy and more neutrons which go on to split other nuclei

Control rods rods – sometimes made from boron – which are inserted into a nuclear reactor to absorb some of the neutrons that would split nuclei, thus controlling the rate at which heat energy is released

Genetic mutations changes to the DNA of an organism which can be caused by exposure to radiation. These changes can often be damaging

Nuclear fission where the nucleus of an atom is split into smaller lighter nuclei and some energy is released

Nuclear power power made from radioactive chemicals, such as uranium

Reactor a device containing nuclear fuel which undergoes fission and releases a large quantity of heat energy and radiation. The reactor is usually surrounded by thick concrete and lead – materials that are good at absorbing radiation

Forces

8

Forces and Newton's Laws

Level 2   What came before?

● SCN 2–07a

By investigating how friction, including air resistance, affects motion, I can suggest ways to improve efficiency in moving objects.

Level 3   What is this chapter about?

● SCN 3–08a

I have collaborated in investigations into the effects of gravity on objects and I can predict what might happen to their weight in different situations on Earth and in space.

Forces and Newton's Laws

Good things come in threes!

No-one has ever seen a **force**. We could say that forces are invisible. However we can all feel forces and we can see what forces do. We all have experience of the *effects* of forces because we live with them every day of our lives. If we stop to think about forces a bit more carefully however, we realise that they can do only **three** basic things.

1 Forces can cause a change in **speed**.

2 Forces can cause a change in **direction.**

3 Forces can cause a change in **shape**.

When two objects collide, forces can affect speed, direction **and** shape at almost the same time.

Sir Isaac Newton derived three Laws of Motion which help us to understand how forces work.

In fact, his laws have become so famous that the unit of force is named after him. Forces are measured in **newtons** (N).

Newton's First Law

Look at the first photograph which follows. The force of gravity, which is actually the weight of the cup, is pulling the cup downwards. However the cup isn't falling. This is because the force in the string is pulling the cup upwards. These two opposing forces must be equal in size and so we say that the forces are balanced.

Now look at the second photograph. This cup is also experiencing balanced forces but there is no string. This time the downward weight of the cup is balanced by an upward force from the table.

In both cases the cups are not moving. They are **stationary**.

This is the **first part** of Newton's First Law. '**Balanced forces** (including zero forces) can allow objects to stay at rest'.

However the Law has a second part to it which is very surprising.

$\Rightarrow$

Aye Aye!

Ask your teacher to show you a toy eye like the one in the following photograph.

If you roll it across a smooth surface it looks as if it is sliding and will never stop.

In fact if an object has balanced forces, or even no forces acting on it, then it can go on and on for ever at a constant speed. That's what the toy eye is doing.

This too would happen to an ice puck after it has been struck by an ice-hockey stick. The puck would slide over the ice and hardly slow down at all.

In deep space a spaceship could be travelling at a very high speed when its rocket engines are switched off. Amazingly the spaceship won't slow down. It will just keep on going at high speed for ever!

This is the **second** part of Newton's First Law. 'Balanced forces (including zero forces) can also cause an object to keep moving in a straight line at a constant speed'.

friction force

engine force

This photograph shows a car travelling at constant speed. This means that the engine force and the friction force are equal in size and acting in opposite directions. (The friction force is made up of *air resistance* and *friction* between the tyres and the road.)

So here, in simple language, are the two parts of Newton's First Law put together:

'If an object is acted upon by balanced forces it will **either** stay at rest **or** keep going in a straight line at constant speed'.

Belt up!

We can see Newton's First Law in action if we consider what goes on inside a car when the driver is forced to make an emergency stop. Engineers deliberately crash test vehicles in order to investigate what happens to the occupants. This helps them to design and build safer cars with improved safety features.

The main safety features in cars today are crumple zones, seat belts and air-bags.

The front of the car is designed to crumple during a collision. The forces involved in a crash therefore change the shape of the car – hopefully, not the shape of the passengers!

2006 LEXUS IS 250
INSURANCE INSTITUTE
FOR HIGHWAY SAFETY
CEF0531

If passengers don't wear a seat belts then when the car comes to a sudden halt, they will obey Newton's First Law! They will carry on travelling at constant speed until they meet something which will exert large forces on them and stop them very quickly. That something is very likely to be a hard surface like the windscreen.

A seat belt provides a smaller less harmful force which will decelerate the passenger much more slowly. This prevents the passenger from hurtling forward to serious injury!

Even when we wear a seat belt we still cannot stop our **heads** from obeying Newton's First Law in a crash situation. This means the driver and front-seat passenger receive facial injuries if their heads continue to move forward and collide with the steering wheel or dashboard. They can also suffer from 'whiplash' which is basically a painful neck injury caused by the head moving forward and then jerking backwards.

Air-bags now help to reduce injury to the face, neck and head by inflating automatically if a collision occurs. Front-seat passengers are protected from facial injury because their heads are decelerated very slowly by a very small force as they sink into the soft air-bag. Also, headrests in the seats help to reduce whiplash injuries when the head jerks backwards.

Even our internal organs have to obey Newton's First Law of Motion! For example, if a person is brought to a sudden halt from a very high speed, his or her heart may continue to move forward inside the body. The heart can be crushed and damaged against the ribs. Engineers haven't found a solution to this problem as yet!

Newton's Second Law

So what happens if the forces acting on an object are not balanced?

Well, quite simply **unbalanced forces** will cause an object to speed up or slow down. Seat belts and air-bags exert unbalanced forces on us to slow us down safely.

A motor car will get faster and faster if the driver presses the accelerator pedal, increasing the pulling force of the engine. Frictional forces will act against the motion of the car but, as long as the engine force is bigger, there will be an unbalanced force acting forward on the car and it will accelerate.

The car in the following photograph is accelerating in the forward direction. The forward engine force is greater than the opposing friction force.

When the driver presses the brake pedal then there will be no pulling force from the engine, but there will be a large unbalanced frictional force acting against the motion of the car. Now the car will slow down.

The car shown in this photograph is decelerating. The brakes provide a large unbalanced frictional force.

Newton's Second Law might therefore be stated as:

'Unbalanced forces cause an object to accelerate or decelerate. The bigger the unbalanced force, the bigger the acceleration or deceleration produced.'

Newton's Third Law

If you shove me, I'll shove you back!

Think carefully here!

Newton's First and Second Laws of motion describe what happens when forces act on **one** object. Newton's Third Law describes what happens when **two** objects interact.

For example, suppose a person pushes against a wall. This force is called an **action force**.

reaction force   action force

The wall doesn't move but the person does … she moves backwards! That's because, when she pushed the wall, the wall pushed her. The person and the wall interact during contact. The force exerted on the person by the wall is called a **reaction force**.

When a bullet is fired from a gun, the gun applies an action force to the bullet but at the same time, the bullet applies an equal and opposite reaction force to the gun. The bullet moves forward and the gun moves backward. The backward movement of the gun is called **recoil**.

A similar thing happens in a jet engine. The jet engine burns fuel and pushes hot gases backwards with an action force. The hot gases push back with a reaction force and the aircraft is propelled forwards.

A toy water rocket works on the same principle.

Be careful! If you do have a water-rocket, then remember to ask a responsible adult, maybe your science teacher, to supervise a launch!

So, Newton's Third Law can be stated as:

'Every action force has an equal and opposite reaction force.'

(Just remember however that the action and reaction forces act on different objects!)

When we drop an apple, it falls towards the Earth because the gravitational pull of the Earth pulls the apple downwards. Suppose this is called the **action force**. Newton's Third Law of Motion tells us that there must be a **reaction force**. The reaction force happens at the same time and the apple pulls the Earth upwards by an *incredibly* small amount!

$\Rightarrow$

Newton's Third Law

Let's go for a twirl!

Here is a very interesting experiment which demonstrates Newton's Third Law.

Take an empty drinks can and punch four holes, evenly spaced out, around the bottom of the can. You can do this with a hammer and nail but be careful! Try to angle the holes slightly and all angled in the same direction. Then tie a length of string, or thread, to the top of the can and, holding the string, lower it into a tank of water, or the kitchen sink.

Let the can fill up with water and then lift it out of the water, keeping it over the tank or sink.

The can spins round quite quickly.

Can you explain how it works?

QUESTIONS

1 State which of Newton's Three Laws of Motion applies mainly to each of the following situations:

   i) A NASA exploration spacecraft travelling to Mars at high speed.

   ii) A pellet being fired from a catapult.

   iii) A person sitting on a chair.

   iv) A car slowing down at traffic lights.

   v) A tennis ball being struck by a tennis racket.

2 Look up the Highway Code, or use the Internet, to find out all you can about 'thinking distance', 'braking distance' and 'stopping distance'.

   Write a paragraph explaining what each of these terms means.

3 a) Read the following paragraph.

   'Force is a **physical quantity** and it is measured in a unit called the **newton**. When a unit is named after a scientist its abbreviated form is always given a capital letter. Therefore, the abbreviation for the newton is **N**.

   Not all physical quantities have units which are named after scientists. For example, **mass** is a physical quantity which is measured in a unit called the **kilogram**. This unit is not named after a scientist and so its abbreviated form is not given a capital letter. The abbreviated form of the kilogram is, therefore, **kg**.'

   Do some personal research and find the names of five scientists who have had units named after them. Make a list of the **Scientists' names**, the **physical quantity**, and the **unit** (make sure you write the unit as a word and also in its abbreviated form).

   b) Now find the units for the following quantities which are not named after scientists.

   Write down the **name** of the quantity and its **unit** in word form and abbreviated form.

   i) Speed

   ii) Acceleration

   iii) Time

   iv) Distance

4 Use the search engine on a computer to find out about the lift force and the drag force which act on an aeroplane wing (known as an aerofoil). To get started, try the Open University website weblab.open.ac.uk.

Numeracy   + − ÷ ×

Newton's Second Law of Motion tells us that unbalanced forces cause objects to accelerate.

This Law also gives us a formula which allows us to calculate the acceleration. The formula is

$$F_{un} = ma$$

Where F_{un} is the Unbalanced Force in newtons,

    m is the mass of the object in kilograms,

    and, a is the acceleration of the object in metres per second per second.

Use this formula to calculate the accelerations of the following objects:

1   a) A family car of mass 1000 kg being pulled by an engine force of 2400 N.

     b) The same family car being pulled by an engine force of 2400 N but also experiencing frictional forces of 600 N.

     c) A toy rocket of mass 0·20 kg launched horizontally from the top of a building by a force of 25 N. (Assume there are no frictional forces at launch.)

2   The World Land Speed Record for a car is held by a vehicle called 'ThrustSSC' which reached a speed of 1228 km/hour, or 341 m/s (763 mph) in October 1997. This is slightly faster than the speed of sound in air and so ThrustSSC had 'broken the sound barrier'. A team of British engineers are working on a project to build a car, called 'BloodhoundSSC', which will reach a speed of 1600 km/hour, or 444 m/s (1000 mph).

The ThrustSSC is a British jet-propelled car. It holds the World Land Speed Record – 1228km/h

The Bloodhound car has a mass of 7 tonnes and will be propelled by a rocket engine which will provide a thrust force of 220 kN.

   a) How many kilograms are in a mass of 1 tonne?

   b) How many newtons are in one kilonewton?

   c) Use Newton's $F_{un} = ma$ formula to calculate the acceleration of Bloodhound.

Useful websites:

You can find out more about Bloodhound on the official website:

www.bloodhoundssc.com

http://www.physicsclassroom.com/class/newtlaws/u2l1a.cfm

GLOSSARY

Balanced forces where two or more forces combine to have no effect on an object

Force the application of energy to cause something to move or change direction

Newton the unit used to measure force, named after Sir Isaac Newton

Stationary not moving

Unbalanced forces where forces have an overall effect on an object

FORCES, ELECTRICITY AND WAVES

Forces

9

Friction and air resistance

Level 2   What came before?

● SCN 2–07a

By investigating how friction, including air resistance, affects motion, I can suggest ways to improve efficiency in moving objects.

Level 3   What is this chapter about?

● SCN 3–07a

By contributing to investigations of energy loss due to friction, I can suggest ways of improving the efficiency of moving systems.

Friction and air resistance

On a cold and frosty morning...

Friction is the name we give to the force which opposes the motion of one object sliding over another.

When your hands are really cold you often rub them together to warm them.

You are actually using the *force of friction* to generate heat energy.

A similar thing happens when a car driver performs an emergency stop and presses the brake pedal. The brake-pads, car wheels and tyres heat up as the surfaces rub together. In these situations, frictional forces are often useful but sometimes they can be a nuisance.

Useful friction

We need friction to help us to do simple things even like walking to school! Without it we would not be able to stay on our feet! Have you slipped on the pavement or in the playground on a freezing cold day? Wintry conditions and lack of friction always result in a sudden increase in the number of people attending hospitals with broken arms and legs caused by slipping on ice!

Formula 1 racing drivers know how important it is to have tyres which generate a useful frictional force. This helps the car to 'grip' the road when travelling at high speed and particularly when taking sharp bends. However most people are surprised to learn that F1 racing car tyres are actually quite smooth! They heat up during the race and become sticky, which then improves grip on the track.

Get a grip!

You can find out just how strong the force of friction can be by trying this 'Gripping Rice' activity.

Use a pencil and a jam-jar shaped container (one with a narrow neck). Add rice until the jar is say $\frac{2}{3}$ full and then use the pencil to make up and down movements in the rice. You will create a depression in the middle of the rice. Pour in some more rice and repeat. Eventually you should feel the rice beginning to grip the pencil. With a little practice you should be able to get the rice to grip the pencil so firmly that you are able lift the jar off the bench!

Also ask your teacher to show you how to do the 'Gripping Books' experiment. It's amazing!

$\Rightarrow$

Get on your bike!

Olympic champion cyclist Chris Hoy likes to ride his bike at top speed but even he has to use the brakes sometimes. The brakes are designed to 'squeeze' the wheel rim and generate a frictional force. This causes the wheel to slow down (and heat up). The same is true of brakes on cars or brakes on trains.

Have you noticed that when we are describing Formula 1 racing cars, bicycles, tyres and so on, the force of friction always acts in the **opposite** direction to the direction of travel. It is usually said that

'Friction always opposes motion'.

Air resistance

Friction can be a real drag

Any two surfaces will produce a frictional force when rubbed together, even if one or more of the 'surfaces' is a gas. Skydivers experience the effects a special frictional force called **air resistance** or **drag** when they are free-falling through the atmosphere. The force of gravity accelerates them to high speeds but as their speed increases, air resistance increases too. A skydiver with his arms tightly tucked in by his sides in a **streamlined** position, called the '**head-down position**', will reach a *steady* speed of about 200 kilometres per hour within 15 seconds. At this speed the air resistance force has increased to the point where it is the same size as the force of gravity (the skydiver's weight)!

The 'head-down' position

The force of gravity (weight) acting downwards is balanced by the force of air friction (drag) pushing upwards.

The skydiver is unable to increase his speed further and so has reached his **terminal velocity** (200 km/h).

He cannot go faster than this speed.

The skydiver can slow himself down however by spreading his arms and legs. This is called the '**belly-to-earth**' **position**. The belly-to-earth position increases the area of his body which rubs against the air and so increases the amount of drag acting against his downward motion.

He will slow down now to a new (lower) terminal velocity of about 120 km/h.

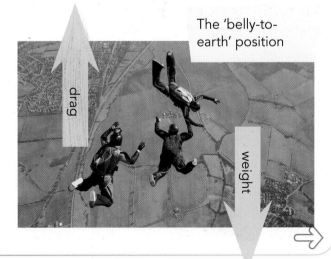

The 'belly-to-earth' position

drag

weight

This lower speed is still too fast for a safe landing however, so skydivers need a way of slowing down to still lower speeds. They do this by opening their parachutes. The parachute is essentially a large area of fabric which greatly increases the drag.

The skydiver slows to a new lower still terminal velocity and descends safely to Earth.

Military jet aircraft sometimes use parachutes to provide extra drag, especially when they have to land on a short runway. The same method of slowing down is used by jet-propelled drag racing cars, or 'dragsters'.

The Space Shuttle

Probably the most spectacular example of how engineers use the force of friction to reduce the speed of a vehicle is that of NASA's Space Shuttle.

When the 100 tonne Space Shuttle is ready to return to Earth from a visit to the International Space Station it has to slow down from a speed of 28 000 km/h in orbit, to a speed of 350 km/h when it touches down on the runway.

Returning to Earth is called **re-entry** and is achieved in stages.

1 First, small thrusters (rocket engines) are used to slow the Shuttle down to 26 000 km/h. This small decrease in speed is just enough to cause it to begin to fall from its orbit.

2 During its time in orbit the Shuttle is actually upside down, so the next job for the thrusters is to flip it over and tilt its nose upward to an angle of 40°.

   It continues to fall from space and enters the Earth's atmosphere at high speed.

3 At this stage the Shuttle experiences drag force from the Earth's upper atmosphere. This frictional drag slows it down to a speed of 10 000 km/h (eight times the speed of sound) and generates a *large* amount of heat energy which could easily destroy the spacecraft. The edges of the wings can reach temperatures as high as 1600° Celsius which is hot enough to melt steel! Special ceramic tiles and other materials on the outside protect the craft and its occupants by absorbing the heat energy.

4 The Shuttle now glides rapidly through the atmosphere in a planned S-shaped flight path to maximise the amount of drag and also to get rid of some heat to the surrounding air. This slows the speed to about 350 km/h.

5 Finally, when the Shuttle lands on the runway, very powerful brakes are applied to the wheels and a parachute opens up to bring the craft to a halt. This takes 70 minutes from when the re-entry operation begins.

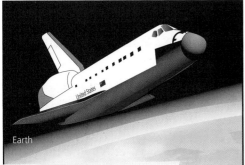

During the early stages of re-entry the nose of the Shuttle is tilted up to 40°

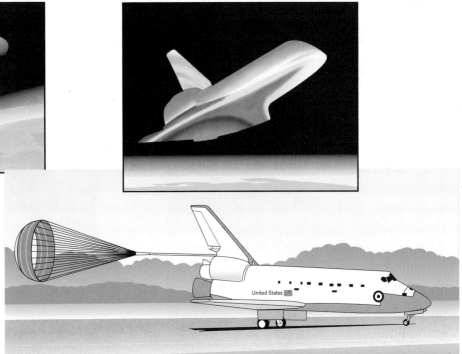

Amazingly, the Shuttle's brakes have to apply such a large frictional force that they reach a higher temperature on the runway than the ceramic tiles do during re-entry!

Not so useful friction – aerodynamics

Yes we need friction to help us to slow things down but can be a real problem when we want to speed things up!

Here on Earth, air resistance always acts to oppose the motion of objects. It is especially obvious for objects which move at high speed. This is why engineers have to think of clever ways to reduce this frictional force. Careful design of the shapes of objects is very important. This is called **streamlining**.

Most cars are streamlined in shape. Racing cars are specially designed so that their curved bodies 'slice' or cut through the air.

The same is true with rockets which are launched into Space.

The Japanese 'Bullet Train' is designed with a streamlined body which helps the train achieve speeds of up to 581 km/h!

Friction and air resistance

Environmentalists criticise Sport Utility Vehicles (SUVs), commonly known as 4 x 4s, because they are relatively fuel-inefficient. The square body shape of SUV means that it encounters a lot of drag and can therefore use as much as 50% more fuel than a typical family car for the same journey!

What do you notice about the shape of this car?

A NASA rocket being launched into space

Hybrid cars – mean machines!

Most of the major car manufacturers are now developing **hybrid** cars. These cars are called hybrids because they use more than one source of energy. Instead of a conventional petrol engine, the hybrid car uses petrol and electricity from batteries to drive an electric motor which then drives the engine. This combination of energy sources is reckoned to be cheaper to run and less harmful to our atmosphere.

Another exciting feature of a hybrid car is its **regenerative braking system**.

The Japanese Shinkansen, or 'Bullet Train', can travel at speeds of up to 581km/h. What do you notice about the shape of the train?

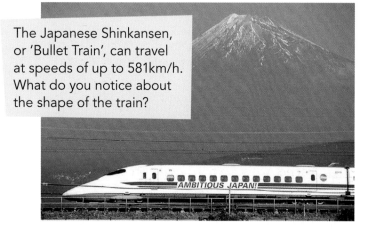

These three examples (F1 car, NASA rocket and Bullet Train) are much more fuel-efficient as a result of their streamlined shapes.

When a conventional petrol-driven car is slowed down or stopped by the brakes, the kinetic energy which the car possessed when it was moving is converted into heat energy in the brake pads. This heat energy is wasted energy because it is lost to the atmosphere.

$\Rightarrow$

Not so useful friction – aerodynamics

When the driver uses the brakes in a hybrid car, the frictional forces are used to operate an electric generator which converts the kinetic energy of the car into electric energy instead of heat energy! This electric energy is in turn transferred back to the batteries. In this way the batteries are being re-charged every time the brakes are used. Less heat is produced, and less energy is wasted!

The batteries still need to be charged up by the car owner occasionally. Instead of going to the petrol station, drivers pull over at charging points at the roadside and plug their cars into a socket!

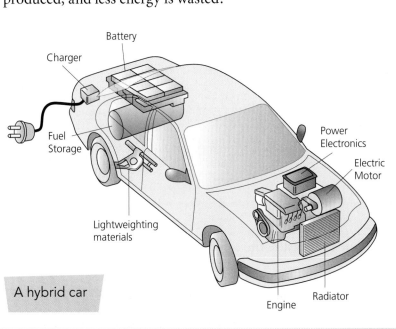

A hybrid car

Floating on air!

The hovercraft is a wonderful example of a machine which has been designed to reduce friction considerably by floating on a cushion of air. Hovercrafts can glide easily over land and water.

You can construct your own model hovercraft with some very simple equipment. You will need a balloon, a discarded CD, some bluetack or plasticine and the top from a plastic drinks bottle. This bottle-top should have a valve which can open and close.

Your teacher will give you advice on building the hovercraft using these items.

Floating on air!

Experimental challenge!

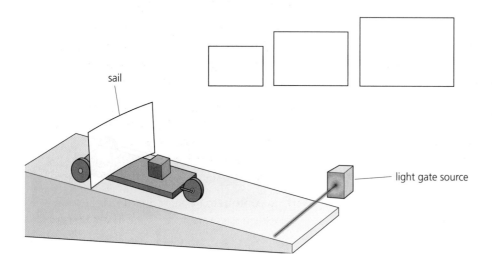

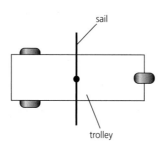

1 Cut several square sails, of different areas, from thin card.

2 Set up a dynamics trolley on a ramp.

3 Arrange a programmed computer and light gate to measure the speed of the trolley at the bottom of the ramp.

4 Attach the sail with the smallest area to the trolley and release the trolley from the top of the ramp. The sail should be at 90° to the direction the trolley is moving (as shown).

5 Repeat the procedure with each sail attached to the trolley and record your results.

6 Plot a graph of trolley speed against area of sail. Plot speed on the y-axis and area of sail on the x-axis.

Additional challenge!

Can you design a shape for a dynamics trolley which will improve its performance on the ramp so that it reaches a greater speed at the bottom?

QUESTIONS

1 Write down the units for (a) Force and (b) Speed. You should write the units in **words** and in the **abbreviated** form.

2 a) Remind yourself what **Newton's First Law of Motion** states about forces. Then use the law to explain briefly why a skydiver reaches a maximum speed (terminal velocity) and cannot go any faster.

   b) Raindrops must have a terminal velocity. Find out the **value** of this terminal velocity. Can you explain why hailstones can hurt us but raindrops don't harm us?

3 Copy and complete the following sentences:

   'The Space Shuttle docks with the International Space Station. They orbit the Earth together at a height of 150km and at a speed of _____ km/h. When the shuttle has to return to Earth, it has to be _____ down. When this is done, it begins to _____ towards the Earth. As it moves through the Earth's atmosphere, the force of _____ causes an energy change to take place. The _____ energy of the spacecraft changes to _____ energy.'

4 a) The Japanese Bullet Train is an example of streamlining. Try to find **six** other examples of streamlining and list them.

   b) Carry out some personal research on **wind tunnels** and how engineers use them to design the shapes of moving vehicles such as of motor cars.

   c) Obtain an image of the International Space Station (ISS) and observe its shape. Why is it not necessary to streamline the ISS even though it is moving at a very high speed?

5 Why do motorists need to ensure that there is an adequate amount of oil in the engines of their cars?

Numeracy $+ - \div \times$

1 The International Space Station (ISS) orbits the Earth at a height of 350 km above the Earth's surface. The **radius** of this orbit is approximately 6750 km measured from the centre of the Earth.

   Calculate the **circumference** of this orbit in kilometres (use the formula $C = 2\pi r$).

2 Write down the formula which connects speed, distance and time.

   Now calculate the **time taken** for the ISS to complete **one** orbit of the Earth assuming the ISS has an orbital speed of 28 000 km/h.

   Now work out how many times the ISS orbits the Earth in one day.

Useful websites:

http://www.care2.com/channels/ecoinfo/hybrid/

http://tryengineering.org/lessons/windtunnels.pdf

GLOSSARY

Air resistance the slowing effect that air has on a moving object

Drag another term for air resistance

Friction force which opposes motion

Hybrid using a combination of engines

Streamlined a shape designed to reduce drag

Terminal velocity the greatest velocity at which an object can fall

FORCES, ELECTRICITY AND WAVES

Forces

10

Don't let gravity get you down!

Level 2   What came before?

● **SCN 2–08a**

I have collaborated in investigations to compare magnetic, electrostatic and gravitational forces and have explored their practical applications.

Level 3   What is this chapter about?

● **SCN 3–08a**

I have collaborated in investigations into the effects of gravity on objects and I can predict what might happen to their weight in different situations on Earth and in space.

Don't let gravity get you down!

Party poppers

Use a party popper on a hard surface like the desk in the classroom.

They can go quite high but what causes them to come back down?

Yes, it's the force of gravity.

Without the force of gravity we would not be able to remain safely 'stuck' to the Earth's surface.

Fall guys

The force of gravity causes all objects to fall towards Earth at the same rate. You can show this for yourself by dropping a large volleyball and a small tennis ball from the same height at the same time. Watch carefully and decide which ball, if any, reaches the ground first! Objects which are compact for their mass are usually seen to fall at the same rate over short distances.

You can see a very famous demonstration of objects falling at the same rate if you look for 'Apollo Mission 15 – Hammer and Feather' on the Internet.

Dave Scott dropping a hammer and a feather on the Moon

Can you explain why a hammer and a feather would not fall at the same rate if released from the same height here on Earth?

Hang about!

The force of gravity acting on an object is called the weight of the object.

You can measure the size of the force of gravity on Earth using a 1 kg mass and a **newton balance**. Simply hang the mass from the hook on the newton balance and measure its weight in newtons (N). This will tell you the size of the force with which the Earth pulls the 1 kg mass downwards.

What is the approximate weight of the 1 kg mass?

Sir Isaac Newton

Hang about!

Different planets and moons in our Solar System have different gravitational field strengths. The 1 kg mass would have a different weight depending on which planet or moon you measured it on!

Use your computer at home or in school to find out the value of the strength of gravity for the Moon, Mars, Venus, Jupiter and Saturn. How do they compare with the strength of Earth's gravity?

If you know your own mass in kg, you should then be able to estimate your weight on the different planets. You should find that on Jupiter you would be so heavy you would barely be able to crawl!

Blast-off!

When the Apollo Space Missions sent American astronauts to the Moon, the spacecraft had to escape from the Earth's gravitational pull. This required a lot of fuel to launch the spacecraft and to help it reach a high 'escape' speed so that it didn't fall back to Earth like a party popper!

Around and around

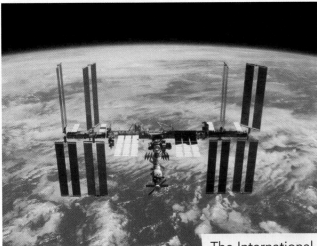

The International Space Station

The Space Shuttle doesn't travel as far as the Moon and so doesn't require to reach such a high speed. It is used to carry supplies and scientists to and from the International Space Station (ISS) which is in *orbit* above the Earth. It is the force of gravity which keeps the ISS and all other satellites in orbit around it and prevents them from drifting off into space!

In our solar system the force of gravity exerted by the Sun ensures that all planets, asteroids and comets stay in orbit around it.

Can you balance the can?

Here is a super demonstration of how to defy gravity!

Try to balance an empty drinks can on its 'edge' on a smooth surface like the kitchen table or a laboratory desk at school. You can't do it – the can just keeps falling over!

Now pour water into the can to a depth of about 3 or 4 mm. You don't need to be exact. Try to balance the can on its edge again. This time, with a little practice, you should be able to do it!

Engineers know that the most stable structures are those which have a low **centre of gravity**. It is easy to topple anything which has most of its mass high above the ground, but very difficult to topple anything which has most of its mass low and close to the ground!

When the can is empty, its centre of gravity is high and it falls over easily. With a small amount of water added, the centre of gravity is much lower and so the can is more stable, making it possible to balance it.

A good example of a tall structure which has a low centre of gravity is The Eiffel Tower in Paris. Its familiar shape ensures that most of its mass is low and close to the ground.

The Eiffel Tower has a low centre of gravity

Can you imagine what would happen if the Eiffel Tower had been built upside-down?

Did you just wink at me?

As engineers design and build structures they often have to find clever ways of overcoming difficult problems. The Millennium Bridge, which spans the River Tyne, was built to allow pedestrians and cyclists to travel easily between Gateshead, on one side of the river, and Newcastle, on the other.

Ships sail up and down the river but the bridge could not be built high above the river or people would have been faced with an exhausting climb from the quayside to the bridge. A pedestrian-friendly low bridge was built with an ingenious way of 'opening up' to allow river traffic to pass through.

Electric motors drive giant underground pistons on both sides of the bridge. These push the base of the bridge outwards and upwards. The large curved beam above the bridge tilts and its weight helps to pull the base upwards.

This additional help from the weight of the beam is called 'counterbalancing' and reduces the amount of energy required from the motors. When the bridge is fully 'open' the force of gravity on the bridge walkway and the force of gravity on the curved beam counterbalance each other and help to hold the bridge in position.

The tilting bridge has been nicknamed 'The Winking Bridge' because, during the tilting operation, when viewed from certain angles, its curved shape resembles an eye closing and opening!

Ebb and flow

The River Tyne in Newcastle, the River Thames in London, and the River Clyde in Glasgow are three examples of **tidal rivers**. This means that if you look at the river in the morning it might be flowing in one direction. If you look at it again in the evening it might be flowing in the opposite direction!

How does that happen? Well, the change of direction is caused by the force of gravity.

The gravitational pulling forces of the Moon and the Sun have a big effect on the directions in which our seas and tidal rivers flow. Because the Moon is much closer to the Earth than the Sun, it pulls the waters in our oceans with a gravitational force which is **twice** as strong as that of the Sun.

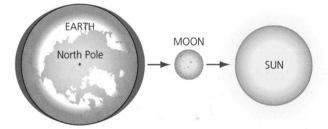

A **spring tide** is caused when the Moon and the Sun are in line with the Earth. The Moon and the Sun can be lined up on either side of the Earth, or they can be aligned on the same side of the Earth as shown. Both situations cause the waters of the Earth to move and produce 'bulges' or high tides on those parts of the Earth which are on that line.

Dumbarton Rock at low tide on the River Clyde

Ebb and flow

Many people think that spring tides occur in springtime, but the name refers to the idea of water 'springing up' to greater heights!

The high and low tides which we see twice daily are a result of the pull of the Moon, and the Earth and the Moon spinning.

Places on the surface of the Earth 'in line' with the Moon experience high tides. Those 'out of line' experience low tides. As the Earth rotates, a low tide location becomes a high tide location when it comes in line with the Moon.

As you are probably aware, there are two high tides and two low tides every day.

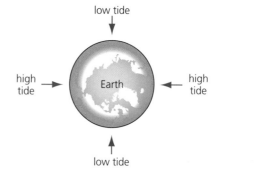

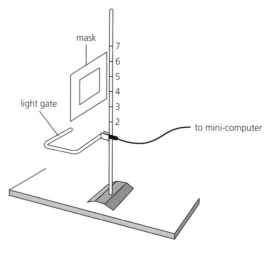

Faster, faster!

When an object is dropped and falls towards the ground it speeds up. This is called **acceleration**. The stronger the force of gravity on a planet, the greater is the acceleration. You can measure the acceleration due to gravity on Earth by doing this simple experiment (you will need help from your teacher to set this up).

Faster, faster!

1 Program the mini-computer to measure acceleration.

2 Drop the mask through the light-gate.

3 The computer will work out the acceleration value for you.

Calculation

Do the experiment at least five times, record all your results, and calculate the average value obtained. Ask your teacher to explain the **unit** of acceleration.

QUESTIONS

1 In what **units** do scientists measure the strength of gravity?

2 Which planet in our solar system has the greatest strength of gravity?

3 The force of gravity makes an object fall towards a planet's surface with an increasing speed. What is this kind of motion called?

4 Copy and complete the following sentence.

'A bag of sugar, of mass 2 kg, weighs ____ newtons on the Earth. On the Moon, the same bag of sugar would have a mass of ____ kg and weigh ____ newtons.'

5 a) A rocket must achieve a minimum velocity in order to 'escape' from a planet's gravitational pull. Try to find the value of this **escape velocity** which a rocket must achieve so that it can leave the Earth and travel into deep space.

b) Try to find the value for the escape velocity from the Moon.

c) Can you explain the difference between this value and the value for Earth?

6 Man-made satellites orbit the Earth.

Find out what is meant by a **geostationary orbit**.

7 Use a search engine on your computer to find out how the Sun, Moon and Earth are aligned when a **neap tide** is produced.

Numeracy   + − ÷ ×

Launch a party popper and use a metre stick to measure the maximum height reached.

Repeat this ten times.

Now calculate the **average height** attained by the party popper after ten launches.

Your teacher may show you how to also calculate the **uncertainty** in your final result.

Useful websites:

www.schoolsobservatory.org.uk

www.howstuffworks.com/question232.htm

http://www2.jpl.nasa.gov/basics/grav/primer.php

GLOSSARY

Acceleration how quickly an object can change its speed

Centre of gravity the point around which an object's mass is balanced

Gravity the name given to the force exerted by one body on another

Weight the force of gravity pulling on an object

FORCES, ELECTRICITY AND WAVES

Forces

11

Buoyancy

Level 2   What came before?

● SCN 2–08b

By investigating floating and sinking of objects in water, I can apply my understanding of buoyancy to solve a practical challenge.

Level 3   What is this chapter about?

● SCN 3–08a

I have collaborated in investigations into the effects of gravity on objects and I can predict what might happen to their weight in different situations on Earth and in space.

Buoyancy

That sinking feeling ...

Why is it that some objects float in water whilst others sink? It is NOT because some objects are heavier than others.

The answer lies with the forces which act on objects when they are placed on the surface or immersed in fluids.

The term *fluid* in science may be used for a liquid or a gas.

Here is an activity you may have done already.

Hang a 1 kg mass from a newton balance and measure its weight. Remember that weight is a **downward** force caused by the pull of gravity.

Now lower the 1 kg mass, still hanging from the newton balance, into a trough of water and measure its weight again.

Wow! The 1 kg mass appears to be about 2N lighter!

This is because the water is exerting an **upwards** force on the 1 kg mass, so that the spring in the newton balance gets a little help. The balance therefore gives a smaller reading.

The upward force which the water exerts on the mass is called **buoyancy**.

Now try this simple experiment:

Blow up a balloon and tie it.

Place the balloon in a trough of water, or the kitchen sink at home.

Press down on the balloon and you will feel a force pushing up against you.

The force you can feel pushing upwards is the force of buoyancy.

When you press the balloon down into the trough you can feel a force pushing up against you

Density

The size of the buoyancy force exerted by a fluid actually depends on something called the **density** of the fluid.

Think of a forest where the trees are planted with lots of space between them. It would be easy to walk between these trees, and we say that the forest has **low density**.

What if the trees were crammed together? It would be really difficult to walk through such a forest.

We would describe the forest then as having **high density**.

$\Rightarrow$

Density

We can use this idea of density to describe the various states of matter.

Some matter, for example gaseous matter, is of low density because its atoms or molecules are far apart, with **lots** of space between them.

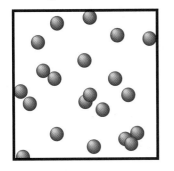

Other matter, in liquid or solid form, is denser because the atoms or molecules are closer together, with **very little** space between them.

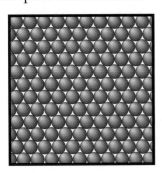

When an object which is made of dense matter is placed in a fluid which is less dense, the upward buoyancy force will be small and it will sink.

So a (dense) brick placed in (less dense) water will sink (like a stone!).

When an object which is made of less dense matter, is placed in a fluid which is more dense, the upward buoyancy force will be large, and it will float. So less dense air-filled balloon floats in water, which is more dense than air.

The brick sinks to the bottom of the water trough

The balloon floats on top of the water

Are you dense or what?

Here are two interesting experiments to try, possibly at home.

Experiment 1 Place an orange in a trough of water. Does it float or sink?

Now peel the orange and replace it in the water. What happens now?

Experiment 2 Place an egg in a large beaker which is $\frac{2}{3}$ full of water. Does it float or sink?

Now use a funnel to pour salty water (brine) slowly into the beaker so that it is guided to the bottom of the beaker. What happens to the egg?

The egg floats on top of the water

Adding salt to water increases the density of the water. The egg sinks in freshwater but, in the more dense saltwater, the egg floats.

Can you explain what is going on with the orange?

The same thing happens when we go swimming. We can float reasonably well if we keep our lungs filled with air and keep paddling with our hands and feet (or wear a lifebelt!). However, in the **Dead Sea,** a person can float easily, because the water is *extremely* salty and much more dense than ordinary sea water.

Practical challenges

Dive! Dive!

This is probably the best buoyancy experiment… **ever**!

It is called the **Cartesian Diver**.

Make your own Cartesian Diver!

All you need is a large (1 litre) plastic drinks bottle and a tomato ketchup sachet.

You must have a sachet which floats **just beneath** the surface when placed in water. If it floats on the surface or sinks to the bottom then don't use it. (You can add blue-tack and/or paper clips to the sachet to get the weight just right.)

1 Push the sachet into the bottle and fill it with water.

2 Replace the bottle-top tightly.

3 Now squeeze the plastic bottle… What happens?

4 Stop squeezing… What happens?

Can you explain what is going on?

Submarines can dive to great depths in the oceans. They do this by pumping sea water into large tanks on board. This increases the weight of the submarine and makes it more dense than the water surrounding it. When the downward weight of the submarine is greater than the upward buoyancy force, the submarine sinks.

In order to surface, the sea water is pumped out of the tanks and is replaced by air. This makes the submarine less dense, the upward buoyancy force is now greater than the downward weight of the submarine, and so it rises.

If there is just enough water on board the submarine so that its downward weight and the upward buoyancy force are equal, then it can 'hover' in the depths below, be very still, and perhaps avoid being detected by enemy ships on the surface.

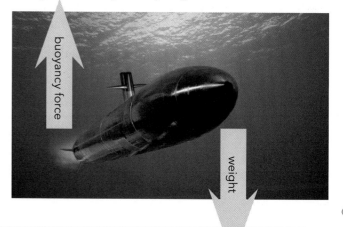

Do try this at home!

Fill a sink with cold water.

Collect a can of a well-known regular soft drink and a can of the diet version of the same drink.

Carefully lower the cans into the water and let them settle.

Note the depth to which each of them sinks (or floats).

Can you explain why some soft drink cans float higher in the water than others?

Once you understand how it works you can test various brands of soft drinks and decide which ones are less fattening!

And another one!

So far, all the ideas about density, floating and sinking have involved gases and liquids.

However here is an interesting demonstration which involves solids!

Find a suitable plastic container. When you buy a stack of CDs they are packed in a clear plastic cylinder which is ideal for this activity.

Fill the container almost to the top with rice. Then place a large dense marble on top of the rice and replace the lid.

Place the container on a bench or table and thump the container firmly several times. Make sure the lid doesn't come off or it gets a bit messy!

It takes a little practice, but you should be able to see through the clear plastic, and observe the dense marble sinking into the rice and disappearing!

Try the same procedure again, but this time **bury** a table-tennis ball in the rice. You should observe the less dense table-tennis ball rising to the surface!

Now you can play a trick on your friends.

You will have buried the table tennis ball in the rice before the trick begins, so that it is unseen. Then wrap the container in paper so that no one can see inside.

Now let your audience see you place the dense marble on top of the rice.

Replace the lid and thump on the bench a few times.

When you remove the lid the marble will magically have changed to a table-tennis ball!

Even solids can sink or float depending on their densities. In fact, the centre of the Earth is made up of dense rocks and metals, such as iron, nickel and gold. Less dense rocks, known as silicates, are to be found nearer the surface of the Earth.

This separation of material occurred mainly when the Earth was being formed approximately 4·5 billion years ago.

Earthquakes and volcanoes have also 'shaken' and 'stirred' the Earth, causing materials in or near the crust of the Earth to migrate up or down.

It is hardly surprising, therefore, that the least dense material in our planet, our **gaseous atmosphere**, floats on top of everything else!

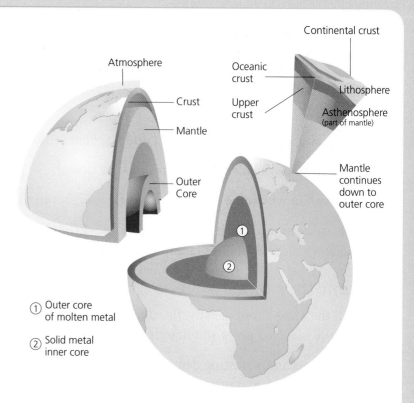

① Outer core of molten metal

② Solid metal inner core

Your very own density tester – stack them up!

Collect small quantities of these four liquids:

Syrup, methylated spirits, water, cooking oil. Now find out what happens when you pour them into the same container. You will need a plastic drinks bottle, a pickle jar, or a laboratory gas jar.

Carefully pour a quantity of each liquid into the jar in any order. A funnel will be helpful.

Let the liquids settle.

Now write down the names of the liquids in order of density starting with the least dense liquid.

Now you can add small objects, such as coins, plastic blocks, polystyrene beads, marbles, and observe how far they sink.

You should now be able to list the materials which you test and place them in a table or list, in order of increasing density.

QUESTIONS

1 Force is a physical quantity which is measured in a unit called the **newton**. Find out what **unit** is used for **density**. You should write down the unit in word form, and in the abbreviated form.

2 The densities of different substances are shown in the table. (The units have been missed out so that question 1 isn't too easy for you!)

Substance	Density
Iron	7870
Zinc	7130
Helium	0.17
Aluminium	2700
Oxygen	1.4
Oil	900

a) Which metal in the table would be best for building aeroplanes? Why?

b) Which metal would be best for making ships' anchors? Why?

c) Explain why a helium balloon rises up when released.

Numeracy $+ - \div \times$

1 Density can be calculated by using the following formula:

$$\text{Density} = \frac{\text{Mass}}{\text{Volume}} \quad \begin{matrix} \leftarrow \text{(in kilograms)} \\ \leftarrow \text{(in cubic metres)} \end{matrix}$$

a) If 2 kg of silver occupies a volume of $0.2 \ m^3$, calculate the density of silver.

b) If 5·2 kg of air occupies a volume of $4 \ m^3$, calculate the density of air.

c) If 3000 kg of fresh water occupies a volume of $3 \ m^3$, calculate the density of water.

d) Using the table of densities given earlier and your answer to part (c) above, describe what you think would happen when oil and water were poured into the same container. Explain what happens when oil is spilled from a ship at sea.

2 The Earth is a sphere.

The volume of a sphere can be calculated by using the formula:

$$V = \frac{4}{3}\pi r^3$$

The radius of the Earth is 6 400 000 metres (6.4×10^6 m).

a) Use the **volume** formula for a sphere to calculate the volume of the Earth.

b) The mass of the Earth is 6 000 000 000 000 000 000 000 000 kg (6×10^{24} kg).

Use the **density** formula to calculate the average density of the Earth.

c) Try to find out the mass and radius of the Moon. Then use the data to calculate the average density of the Moon.

Numeracy + − ÷ ×

Numeracy extension calculation

Did you know that if we could find a container of water large enough, we could drop the giant planet Saturn into it and it would float?

If you can find mass and radius data for Saturn, then calculate its density.

You should be able to explain why Saturn would float in water!

Saturn is the second largest planet in the solar system

Useful websites:

To find out more about submarines try,

http://science.howstuffworks.com/submarine.htm

http://en.wikipedia.org/wiki/submarine

GLOSSARY

Buoyancy the force exerted on an object when it is placed in water

Density the amount of mass in a certain volume

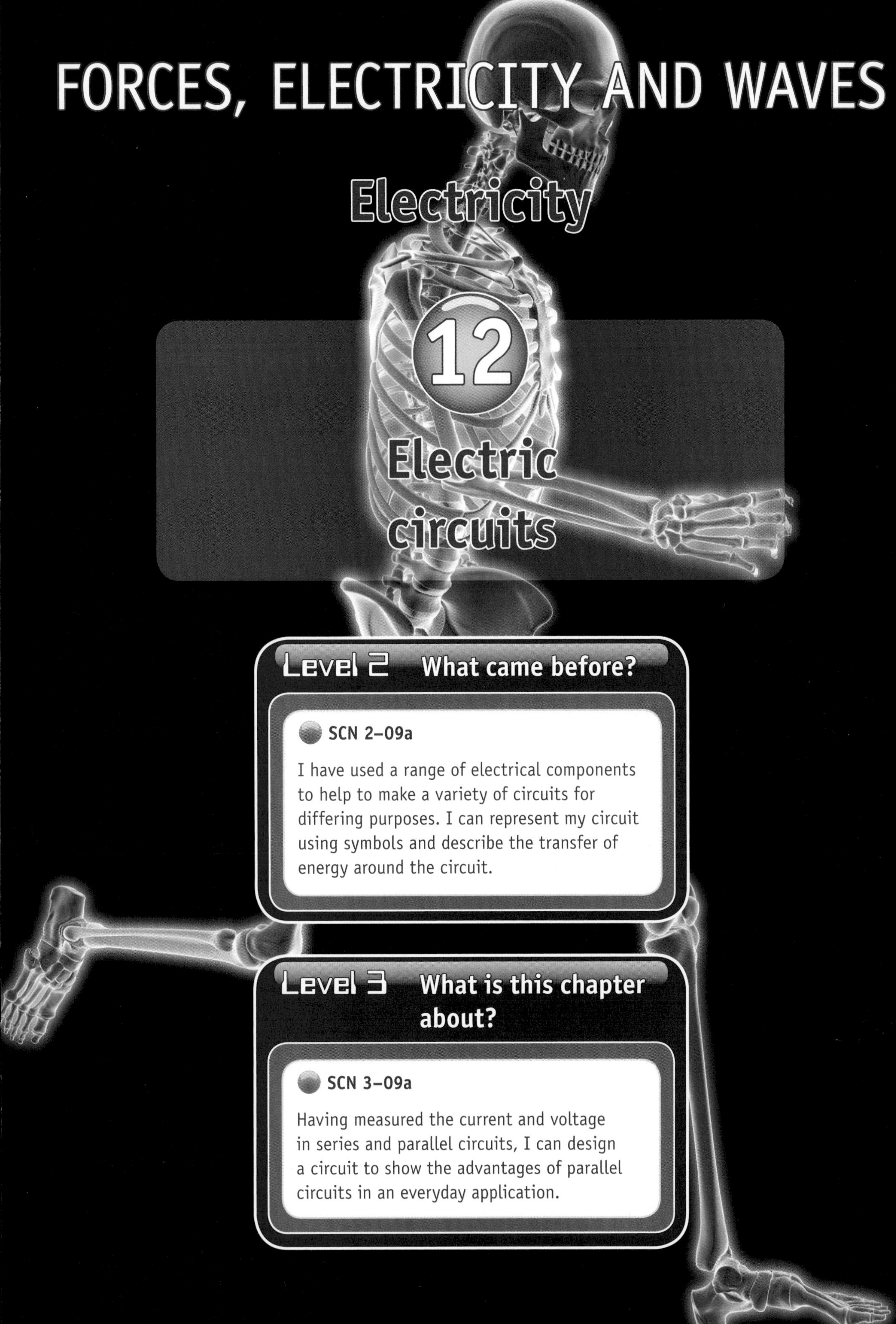

FORCES, ELECTRICITY AND WAVES

Electricity

12

Electric circuits

Level 2 — What came before?

SCN 2–09a

I have used a range of electrical components to help to make a variety of circuits for differing purposes. I can represent my circuit using symbols and describe the transfer of energy around the circuit.

Level 3 — What is this chapter about?

SCN 3–09a

Having measured the current and voltage in series and parallel circuits, I can design a circuit to show the advantages of parallel circuits in an everyday application.

Electric circuits

Electricity is the general name we give to the subject of electrical energy, how it operates and how it is transferred. The word 'electricity' comes from the Greek word for amber 'electros'. In trying to understand electric **circuits** it is important NOT to use the word 'electricity' when explaining or describing how a circuit is used. Many people and textbooks use phrases like:

'the electricity travels from the battery to the light bulb'

'the electricity causes the wire to heat up'.

The problem arises when people interpret the word 'electricity' as energy, or charge, or power, or current. This leads to people reading about or explaining the same event but understanding it in quite different ways!

Background knowledge

You should already be familiar with simple circuits, symbols and how we use them.

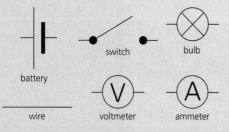

We use the symbols for clarity. The circuit diagram which follows represents the actual circuit shown.

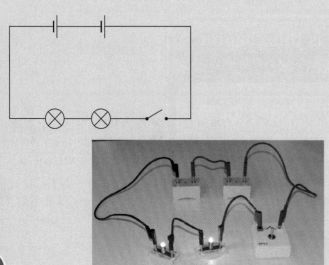

The key to explaining how this circuit operates relies on these principles:

- It needs a supply or source of electrical energy.

- It needs to be a *complete* circuit of conducting materials (conductors). Most electrical cables and wires are made from copper. Copper is a metal which has electrons that move around easily.

A good answer to the question 'How does this circuit work?' could be:

'When the switch is closed, it completes the circuit. This complete circuit allows energy to travel from the battery to the bulbs and they light up.'

Looking at the circuit, what would happen if a lead was disconnected from one of the bulbs?

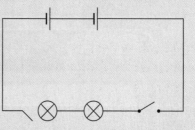

Obviously *both* lights go out.

Why does this happen?

A good answer could be:

'When a lead is removed this breaks the circuit. If there isn't a complete circuit the energy cannot travel and the lights go out.'

These are not the only ways of answering questions like these but they are simple and straightforward. In circuits we cannot see things like current or charges directly so it is very easy to misunderstand what is actually happening. When this subject was being studied initially by scientists it was poorly explained for many years.

Series and parallel circuits

There are many different ways of combining electrical components so that they operate in the way we want them to.

A **series circuit**, like the ones shown on page 98, has the advantage that when we switch it on, both lights come on at the same time. Both lights go off at the same time too.

A disadvantage is that if one bulb was to fail, the other would not light.

It is called a series circuit because the components are connected so that one follows the other. They are connected 'in series'.

Another way of connecting the same components would be like the circuit to the right.

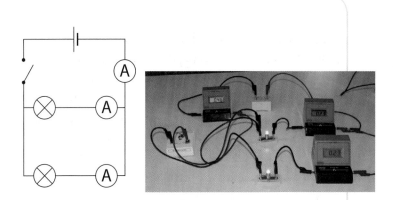

This is referred to as a **parallel circuit** since the two bulbs are shown one below the other. They are parallel to each other. They are connected 'in parallel'.

Current in a series circuit

In the earlier explanation of how a series circuit operates: 'When the switch is closed, it completes the circuit. This complete circuit allows energy to travel from the battery to the bulbs and they light up', no specialised scientific terms were used.

We can now look at this in more detail and use special terms to give a more sophisticated explanation.

When a bulb is connected to a battery it lights up. The battery produces small electrically charged particles called electrons. These electrons move in the circuit to make it operate.

The movement of these electrons in the wires is called an electric **current** and is measured in *Amperes* [named after a French Scientist called André-Marie Ampère (1775–1836)]. We shorten the word Ampere to Amp and use the letter 'A' for the unit. So 10 Amperes could be written as 10 Amps or 10 A.

We measure current using an *Ammeter*

The symbol for an ammeter

An ammeter is connected in *series* in the circuit.

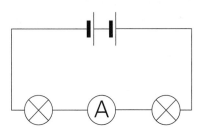

It is the movement of electrons in a wire that is in fact the current. When more electrons move in the wire the current is larger. The current in the cable of a kettle might be around 8 Amps. The current in the cable of a radio might only be around 0.2 Amps.

Electric circuits

More powerful appliances need larger currents to operate.

In the last circuit we measured the current in the wires between the two bulbs.

What do you think the current will be at other sections of the circuit? More, less, the same?

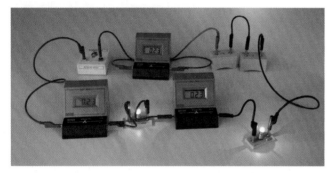

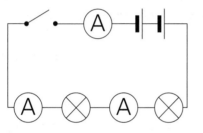

As can be seen, the current in a series circuit is the same in all sections. This is because a series circuit only has one route for the electrons to move along.

If we had 500 electrons leaving one side of the battery, we would need to have 500 electrons moving in the bulb and 500 electrons moving towards the battery. We couldn't 'lose' any along the way because there are no other connectors for the electrons to move along.

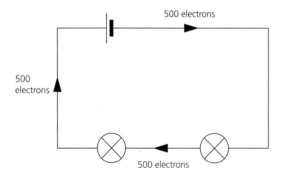

500 electrons

500 electrons

500 electrons

Now you may be more familiar with how the current in a series circuit acts. It is the same in all sections. You can also now explain the circuit in more detail.

The answer given earlier to the question 'How does this circuit work?' was:

'When the switch is closed, it completes the circuit. This complete circuit allows energy to travel from the battery to the bulbs and they light up.'

This could now be changed to:

'When the switch is closed it completes the circuit. A current flows through the connectors and this causes the bulbs to light up.'

Both answers are correct but the second one has a little more scientific terminology.

Series circuits are reasonably straightforward but have the disadvantage that a break in any part of the circuit means that it is not complete, and therefore all appliances in that circuit go off at the same time.

This is a major problem. Could you imagine if all your appliances at home were connected in series? Everything, for example cooker, TV, computer would have to be on or off at the same time! This isn't really workable so we must use other circuits which allow us to switch things on and off individually.

Current in a parallel circuit

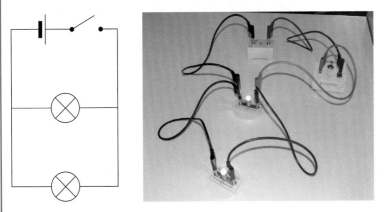

As shown here, the current in a parallel circuit is **not** the same in all sections. This is because there are different paths for the electrons to travel. There is a special rule for the current in parallel circuits, which you might have guessed already from the ammeter readings shown.

The current from the supply is equal to the sum of the currents in the individual sections.

Mathematically $I_s = I_1 + I_2$. We use 'I' for current. In the equation I_s is the current from the supply; I_1 and I_2 are the currents in the sections.

It is straightforward to explain. The current from the battery can 'split'. Some of the electrons travel along one route of the circuit and some travel along the other. The currents then 'rejoin' when the two sections rejoin.

Parallel circuits are more useful than series ones because they give us the ability to switch on appliances independently of each other. The lights in your homes are in a type of parallel circuit. This means you can switch the lights on and off individually.

The currents that appliances use can vary a great deal. A hair dryer may use 6 A but an MP3 player will use less than 0.1 A.

Starter motors in cars use incredibly large currents when the ignition is turned. These currents can be as large as 100 A!

To protect the wiring in the home from overheating, we use fuses and circuit breakers at various points. The plug for your toaster will have a fuse rated at 13 A. If the current goes beyond 13 Amps for a period of time, this fuse will melt (the old word for 'melting' was 'fusing') and disconnect the toaster. You would then have to repair any fault and replace the fuse for it to work again.

The wiring in your home will have a 'fuse box' or more likely a box of circuit breakers. When the current from a room or from a range of appliances gets too large it will disconnect the circuit and so not allow the wires to overheat. You then have to replace the fuse or reset the circuit breaker. These safety devices ensure that electrical faults no longer tend to cause house fires.

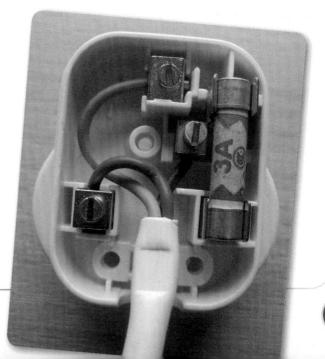

Voltage in a series circuit

The term **voltage** is referred to in many circuits and denoted by the letter 'V'. Our mains electricity supply is rated at 230 V. We can buy batteries at 1.5 V or 6 V or whatever is required.

The term voltage refers to the *work done* by the electric charges when they operate in a circuit.

This concept is extremely difficult to understand and it is better to observe how voltages behave in circuits. This then gives us a clearer idea of what voltage is all about.

(The scientist Volta investigated the relationship between electric charge and its effect on how circuits worked and used energy. The emperor Napoleon so liked the idea of such an impressive scientist working for him, that he made him a Count and reduced his teaching at the University to one lecture a year!)

We measure voltage using a Voltmeter.

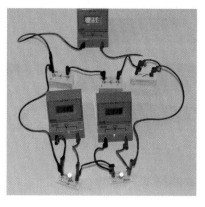 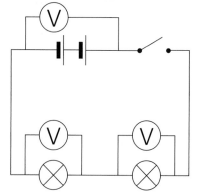

The symbol for a voltmeter

We can see that in this series circuit the voltage from the battery is 'spread' between the components in the circuit. We say that the supply voltage is equal to the sum of the voltages across the components in the series circuit.

Mathematically $V_s = V_1 + V_2$

In this equation V_s is the supply voltage; V_1, V_2 are the voltages across individual components.

These circuit diagrams show further the splitting of voltage in a series circuit.

In series circuits therefore we have two rules:

1 The current in the series circuit is the same at all points in the circuit.

2 The voltage from the supply is split between all the components in the series circuit.

If you can remember these rules you have a good knowledge of how a series circuit operates.

Voltage in a parallel circuit

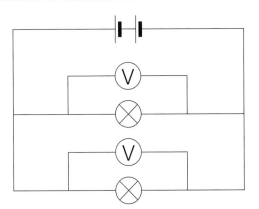

This circuit diagram shows that a parallel circuit behaves quite differently from a series circuit.

The voltage of the supply is the same as the voltage across each of the parallel components.

This is because a parallel circuit is like two separate circuits connected to the same power supply. In effect, each branch of a parallel circuit acts independently of the other. This means that the voltage in each branch is the same as the supply or battery voltage.

If we were to connect say five bulbs in parallel with a battery, all bulbs would receive the same voltage. This would drain the battery very quickly however.

In parallel circuits therefore we have two rules:

1 The currents in parallel circuit branches add to give the current taken from the supply.

2 The voltage in parallel circuits is the same for all components in parallel.

Useful circuits

Our homes have many electric circuits.

The electrical energy we require comes from power stations via the National Grid. The cables in our homes are connected in such a way that we can use them easily and safely. Most of the circuits in the home are variations of parallel circuits.

An example of this is a multi-block plug extension.

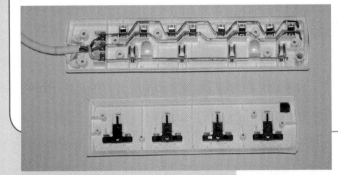

These are connected in parallel.

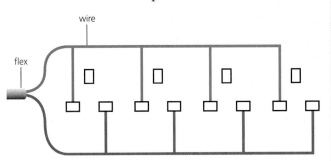

If you look at the way the wires are connected to the extension block you can see that when we insert plugs they are connected 'in parallel'. This allows us to add or remove appliances to it without them interfering with each other.

GLOSSARY

Circuit a combination of connectors, components and power supply

Current the flow of charged particles in a wire

Parallel circuit a circuit where the connectors form branches and components can be connected to separate branches

Series circuit a circuit where each component is connected one after the other

Voltage a measure of the energy available in a circuit

FORCES, ELECTRICITY AND WAVES

Electricity

13

Batteries and cells

Level 2   What came before?

● SCN 2–10a

To begin to understand how batteries work, I can help build simple chemical cells using readily-available materials which can be used to make an appliance work.

Level 3   What is this chapter about?

● SCN 3–10a

I can help to design simple chemical cells and use them to investigate the factors which affect the voltage produced.

Batteries and cells

Previously we saw examples of simple circuits and how we combine them in series and parallel to design other circuits to power our homes, workplaces and streets.

Many of our modern electrical and electronic devices are portable. We carry them about freely so that our everyday life is made easier.

These devices all require some sort of electrical power supply in order to work and that means using batteries.

As can be seen the range and type of batteries today is vast. They range in size and energy requirements dramatically.

We have cars that are electrically powered which match the performance of petrol cars. Batteries are also important for reducing car pollution in the long term.

Beyond our need for portable supplies of electrical energy batteries can play a significant and vital part in our use of renewable energy sources. Batteries have some advantages over mains electricity as they produce lower voltages, which make them safer to use.

Many renewable energy sources generate energy when the conditions are correct, but not necessarily when we need the energy. This means we need to store the energy when it is being generated and use it when required. Batteries can be used to store this surplus energy.

When a battery is working there is an energy change taking place. Chemical potential energy inside the cell is changing into electric energy. The electric energy produced is a result of a chemical reaction. However, like all chemical reactions, the reaction slows down as the chemicals are used up. Finally with all the chemicals spent, the battery stops working all together.

Frog tales and sour electricity

In 1749 the American Benjamin Franklin first used the word '**battery**' to describe a number of glass jar *capacitors*, which could store electrical charge. The stored charge could be released as an uncontrolled, but spectacular, and sometimes dangerous spark. If the jars were left unused for too long it was found they had no 'spark'. The stored energy would somehow 'leak' away.

The first true battery to give a controlled current was produced in 1800 by an Italian scientist called Alessandro Volta. Volta's home was near Lake Como in northern Italy, which is now a popular tourist destination. There is a museum there, the Voltian Temple, which still has some of Volta's equipment.

$\Longrightarrow$

Volta made an electro-chemical cell from copper and zinc metals separated by strong paper soaked in salt and water. He took inspiration from one of his contemporaries Luigi Galvani who thought that a dead frog's leg could be made to twitch when touched by an iron probe, as long as the frog was attached to a brass hook. He believed electric energy caused these muscle contractions.

The key idea was that it required two different metals, slightly apart, but connected by contact with some liquid. In the example of the dead frog, it was its body fluids that connected the metals.

Volta used salty water (brine) to connect the metals.

This connecting liquid is known as the **electrolyte**. The two metals used by Volta, copper and zinc, are known as the **electrodes**. Volta arranged his cells, one on top of another, to form what is now known as a **Voltaic pile**. Unfortunately, if the pile became too tall, the weight of the cells squeezed the electrolyte out, and the pile no longer worked. This limited the **voltage** Volta could produce from his invention.

The problem of the electrolyte being squeezed out by the weight of the metals pushing down was solved by an Englishman named William Cruickshank. He stacked the metal plates and electrolyte horizontally in a wooden box. This arrangement became known as the trough battery.

Cells arranged or stacked together in a pile are called a battery but for most people the words 'cell' and 'battery' are interchangeable and just refer to a portable source of electric energy.

How to make a cell like Allesandro Volta's

Take 1 lemon and roll it about on the bench to 'free' the juice inside it.

Take one clean piece of zinc (a galvanised roof nail would do) and one clean piece of copper (a clean copper coin).

You will also need a voltmeter.

Insert the zinc and copper (the electrodes) deep into the lemon, making sure that they do not touch.

Connect a voltmeter between the electrodes and record the voltage you have produced.

If we connect a few of these in series then the voltage is increased.

Why does the lemon cell produce electrical energy?

Batteries like this work because of chemical reactions between the different metals and the electrolyte connecting them. Metals react with the electrolyte in different ways and if we choose the combination of metals correctly we can get larger or smaller voltages produced.

This topic is known as **electrochemistry** and metals can be arranged in the way they react with electrolytes.

This arrangement leads to the **electrochemical series** and it places the metals in order of how much voltage they can produce under standard conditions.

Zinc is higher in the electrochemical series than copper so it loses electrons more readily than copper when put in a solution. Zinc metal forms **zinc ions** when put in the lemon juice. This loss of electrons is called **oxidation**. It may be described by an *ionic equation*:

$$Zn \rightarrow Zn^{2+} + 2\ e^-$$

The zinc ions (Zn^{2+}) move into the lemon juice. The lemon juice forms a conducting link between the zinc and the copper electrodes and the circuit is completed. The zinc metal now has lots of 'free' electrons on its surface.

These extra electrons on the zinc metal flow from the zinc to the copper through the wires connecting the electrodes.

This is what causes the electrical energy to flow and gives us the 'voltage'.

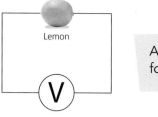
Lemon

A circuit diagram for a lemon cell

Improving the cell

With early batteries there were some problems. At the copper electrode, hydrogen ions in the lemon juice gained electrons to form hydrogen gas bubbles. This hydrogen gas affected the battery quite badly.

Thirty-six years after Volta produced his voltaic pile, a British chemist, John Daniell found a way to prevent hydrogen gas building up. Daniell used two electrolytes instead of one. The second electrolyte absorbed the hydrogen gas.

With this arrangement (conducting) copper metal was produced instead of hydrogen gas. The Daniell cell had an operating voltage of roughly 1.1 volts and was widely used in the early telegraph networks.

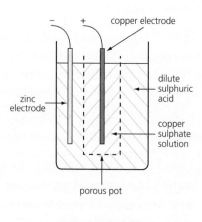
copper electrode

dilute sulphuric acid

zinc electrode

copper sulphate solution

porous pot

A Daniell cell

Battery voltage and current

The voltage output of a cell is a complex matter. The main factor is the electrochemical difference of the metals in the cell. The surface area of the metals and the strength of the electrolyte determine mainly the time for which the cell can run.

To increase the voltage to operate different devices it is easy to combine cells in series or parallel in much the same way as bulbs were connected in series or parallel.

When cells are arranged in series, their voltages add up.

For example, six 1.5 V cells in series give a voltage output of six times 1.5 V which is 9 V.

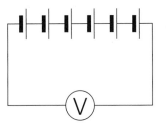

This is seen in devices like torches and remote controls where we need a voltage of about 6 volts. We simply combine the cells in series and connect them so that they act as one large battery.

How long a battery will last is difficult to say. It depends upon how it is used. In general terms if a large current is used the battery will last for a short time. If a small current is used the battery may last for a long time.

To get an estimate of how long the battery will operate for we use the **amp-hour** rating (**Ah**). The amp-hour (Ah) rating allows us to calculate how long the battery will be able to perform a certain function. Battery manufacturers state the maximum current to be drawn from a battery along with its amp-hour rating.

For example, if you had a 1.5 V, 0.5 amp-hour battery, an output voltage of 1.5 V would be possible along with a current of 0.5 A for 1 hour.

If we placed the battery in a circuit where a current of 0.25 A was drawn then we could expect the battery to operate for 2 hours.

Numeracy + − ÷ ×

1 What would be the voltage and current if a 1.5 V 0.5 Ah battery was used to provide 1.5 V in:

   a) 0.1 A circuit?

   b) 0.75 A circuit?

   c) 1.5 A circuit?

2 Electric wheel chairs typically use two 12 V lead acid batteries in series. What voltage would be supplied to the motor?

Battery types

The range of batteries available today is great. While they operate on the same basic principle described they are designed for many uses.

Some devices use very little current and simple and cheap batteries are the best and most cost-effective way of powering them. Remote controls for TVs and CD players use such batteries.

Some devices, such as a car a battery is required to provide a very large current for a short period of time. It can then be recharged once the car has started running.

Other devices need to provide a small but reasonable supply of energy that can last for a day or so, but can also be recharged for use again the following day. Laptop computers, mobile telephones and MP3 or MP4 players fall into this category.

We generally categorise batteries into **primary** or **secondary** types.

Primary batteries are for single use and can be disposed of once depleted.

Secondary batteries are rechargeable and designed to be used over and over again.

The following table provides a useful list of most of the batteries available today.

Battery type	Battery class	Uses	Comments
Zinc-carbon	Primary (dry cell)	Toys, torches, remote controls, radios, clocks, smoke alarms	1 Zinc and carbon electrodes with an acidic electrolyte. 2 Cheap and has a tendency to leak.
Alkaline-manganese	Primary (dry cell)	As above	1 Zinc powder and manganese oxide electrodes, with an alkaline electrolyte. 2 Longer lasting and more expensive than zinc-carbon, although of a similar size.
Zinc-air	Primary (button)	Hearing aids and experimental vehicles	Energy is produced as zinc is oxidised with oxygen from the air.
Silver-zinc	Primary (button)	Watches, aircraft, weaponry and the Apollo space missions	As silver is expensive this battery is only used when cost is not a concern.
Mercury oxide	Primary (button)	Early pacemakers, calculators, hearing aids, photographic light meters	Has a very steady voltage over a long life span. Now banned and being replaced by silver-zinc.
Nickel-cadmium (Ni-Cd)	Secondary (dry cell)	Cordless power tools, shavers, emergency lighting and remote controlled toys	1 Cadmium metal and nickel oxide hydroxide electrodes. The first rechargeable battery. 2 Copes well with power surges.
Nickel-Metal hydride (Ni-MH)	Secondary (dry cell)	Portable electronics – computers, mobile phones	1 A hydrogen absorbing alloy replaces the cadmium metal electrode. 2 They are more environmentally friendly than Ni-Cd. 3 They have 2 to 3 times the capacity of Ni-Cd.
Lithium-ion (Li-ion)	Secondary (dry cell)	Mobile phones, iPods and laptops, consumer electronics, defence, aerospace and automotive application	Lithium ions move between the electrodes.
Lead-acid	Secondary (wet cell)	Car batteries, fork lift trucks, golf trolleys, emergency lighting and uninterrupted power supplies	1 The acid used is sulphuric acid. 2 The battery is rechargeable but very heavy. 3 Can provide high current but loses its charge over a few weeks.

QUESTIONS

1 Give two advantages that batteries have over mains electricity.

2 What energy change takes place inside a battery when it is operating?

3 What components form a simple electrochemical cell?

4 What is the purpose of the electrolyte in a cell?

5 Why does a battery eventually stop working?

Battery problems

Waste

We all use battery-operated devices each day but there are problems with this, some of which can be quite serious. One serious problem is waste. In Britain each year we dispose of approximately 20 000 tonnes of general purpose batteries, and over 110 000 tonnes of car and van batteries. Less than 2% of single use batteries are recycled by the public. The rest are dumped in landfill sites. Britain has a target that by 2016, 45% of batteries used will be recycled.

Suppliers of batteries will have to do their bit too. From the 1st February 2010, shops and stores who sell more than 32 kg of batteries a year will need to accept used portable batteries from consumers, without charge.

Toxic matters

Batteries are classed as hazardous waste, and may only go to landfill, if the site has a special license for this waste type. There are very few such licensed landfill sites in Scotland.

Batteries contain hazardous chemicals which leak out over time. The chemicals combine with the rainwater running through the landfill site and the liquid run-off can enter nearby streams and rivers. When such contaminated rainwater reaches nearby water courses, it can increase the concentration of certain 'red list' substances in the environment.

Mercury and cadmium which are found in batteries are examples of 'red list' (List 1) substances. These are chemical substances which cannot be broken down by nature, but instead accumulate in food chains where they can lead to abnormalities and even death in some species.

With the *Waste Batteries (Scotland) Regulations 2009*, it will be illegal to landfill or incinerate industrial or automotive batteries after 1st January 2010. In addition, legislation will ensure the following from October 2011:

- 65% recycling of lead acid batteries by weight

- 75% recycling of nickel-cadmium batteries by weight

- 50% recycling of other types of batteries by weight.

Battery fires

There have been a number of reports in the media about laptops and iPods going on fire!

Laptop computers generate a lot of heat during normal working. This heat can affect their performance and battery operation quite badly.

The general advice for a laptop is **not** to sit it on your lap for long periods.

Batteries should not be stored amongst metal objects, which can 'short' their terminals.

Steel wool, coins and odd pieces of wire should never be thrown on top of batteries, in your pocket or in a storage box.

Remember that all fires need a source of heat, and given the right conditions a battery can start a fire, especially if there are other flammable materials nearby.

GLOSSARY

Amp-hour an indicator of the amount of time for which a battery can operate

Battery a portable supply of electrical energy

Electrochemical series a list which ranks elements by their ease to form ions when put in solution

Electrochemistry a branch of chemistry which studies the chemical reactions between electrolytes and electrodes

Electrode a conducting fluid used to connect metals in a battery (electrodes)

Oxidation where a metal atom releases electrons

Primary batteries single-use batteries which are disposed of after use

Secondary batteries rechargeable batteries which can be re-used

Voltage a measure of the energy available in a circuit

FORCES, ELECTRICITY AND WAVES

Waves

14

Pushing the boundaries

Level 2   What came before?

● SCN 2–11b

By exploring reflections, the formation of shadows and the mixing of coloured lights, I can use my knowledge of the properties of light to show how it can be used in a creative way.

Level 3   What is this chapter about?

● SCN 3–11a

By exploring the refraction of light when passed through different materials, lenses and prisms, I can explain how light can be used in a variety of applications.

Pushing the boundaries

In our earlier work on the Electromagnetic Spectrum (Chapter 1), we looked at the nature of light and its relation to other electromagnetic waves. In this chapter and the next we look at the behaviour of light.

Reflection and refraction

The behaviour of light waves at a boundary between two transparent materials is of great interest when designing optical systems. Optical systems, such as microscopes and telescopes, are designed to control and make use of light. These optical systems use specially shaped **lenses**, similar to spectacles, which help us to focus on near or distant objects when our eyes are not quite up to the job.

When light energy arrives at the boundary between two transparent materials like air and water, or air and glass, some of the light energy is reflected. This means that some of the light 'bounces off' the second material without entering it. This reflected light obeys the **law of reflection** with which you should already be familiar.

In this chapter we look at what happens to the light energy which is **not** reflected, but **enters** the material. Light which leaves one transparent material and enters a second transparent material is said to have undergone a process called **refraction**.

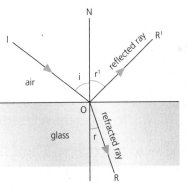

A closer look at refraction

1   The amount of incoming light energy which is reflected at the boundary and the amount of light energy which enters at the boundary all depends on the two materials forming the boundary.

Consider a boundary between air and glass, both of which are transparent. Some of the light is reflected off the surface of the glass. Some of the light enters the glass and does not reflect. This ray of light undergoes refraction and alters its direction and speed.

Your teacher can help you to set up a simple physics experiment to investigate what happens to a ray of light from a ray-box when it enters a glass block. The following diagrams may help.

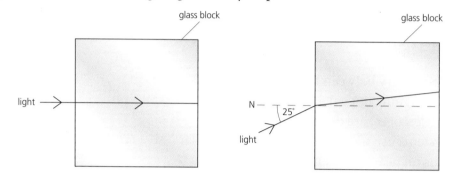

You should notice that, as you increase the angle at which the light ray approaches the air–glass boundary, there is an obvious change in direction of the **refracted ray** in the glass.

(Your teachers may supply perspex blocks instead of glass. Perspex refracts light in much the same way as glass and will produce very similar results. It doesn't break quite as easily!)

Why do you think this change in the direction of light occurs? Historically it caused much interest and debate.

The answer to the question is this:

When light travels through air it travels at a speed of almost 3×10^8 m/s. When it enters another transparent material such as glass, however it is slowed down.

In glass, its speed is reduced to about 2×10^8 m/s.

It is this **change in speed** which causes the **refraction** of light.

2 Look at the two diagrams which follow. These show a ray of laser light passing from air into glass and another ray passing from air into water. The rays have the same angle of incidence.

Reflection and refraction

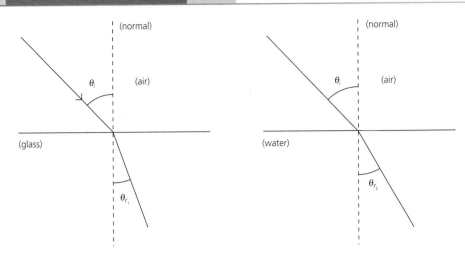

Notice that a very important dotted line has been added in each diagram. This line is drawn at right angles to the boundary between the material and is called the **normal**.

The normal has to be shown in the diagrams to allow the **angle of incidence** (θ_i) and the **angles of refraction** (θ_{r_1}, θ_{r_2}) to be marked.

Notice how these angles are measured between **ray** and **normal**.

In each diagram the incoming, or **incident**, ray approaches the boundary at the same angle. However, the glass causes the light ray to change direction by a greater amount than the water. (Look at the **refracted** ray in each diagram.)

The reason for this is that glass is **more dense** than water.

The **density** of a transparent material is the main factor which controls the amount by which a light ray will be slowed down.

Light rays travel through air at a speed of about 3×10^8 m/s, but glass will reduce the speed of light to 2×10^8 m/s. Water will slow down light rays from 3×10^8 m/s to about $2 \cdot 3 \times 10^8$ m/s.

The greater the density of the transparent material, the more it will reduce the speed of light. Glass is denser than water, light therefore travels slower through glass than it does through water, and so glass refracts light more than water.

Look up the Index!

Here is a very simple formula which can be used to indicate how much refraction will occur when a ray of light travels into a transparent material.

$$n = \frac{\text{Speed of light in vacuum (or air)}}{\text{Speed of light in transparent material}}$$

The symbol n is called the **refractive index** of the material.

If we use a speed of 3×10^8 m/s in air, and a speed of 2×10^8 m/s in glass, then the refractive index of glass works out to be 1.5 (Check it out with a calculator!).

Glass comes in lots of different grades and depending on the density of the grade, it can have a range of refractive indices. Crown glass has a refractive index of 1.48 but denser flint glass, which contains lead, can have a refractive index as high as 1.80.

Can you use the formula to calculate the refractive index of water? The speed of light in water is 2.25×10^8 m/s^{-1}.

Experiment

Set up a ray box and rectangular glass block on top of a sheet of white paper.

Trace an accurate diagram which shows the shape of the block and the path of a single ray of light (incident from air at about 40°) as it passes through the glass and leaves the other side of the block. (Ignore any faint reflected rays which may be visible at each boundary.)

Then use a protractor and ruler to draw in the normal at each boundaries.

Measure the angles of incidence and refraction at the boundary where the light ray **enters** the glass. Then measure the angles of incidence and refraction when the light ray **exits** the glass.

(You should seek help from your teacher if necessary.)

What conclusions can you reach?

The diagram shows us that, when light travels from one medium (air) into a denser medium (glass), the light ray will change direction **towards** the normal. (Remember too that the light ray is slowing down.)

However at the second boundary, when the light ray travels from a denser medium (glass) into a less dense medium (air) and the light ray bends **away from** the normal. (Remember too that the light ray is speeding up.)

Into reverse!

Imagine the same diagram but with the direction arrows turned around. This is the path the ray of light would take in the reverse direction.

All types of electromagnetic radiation obey this law, which is known as the **law of reversibility**.

The law states that if the direction of the light ray was reversed the new path would lie exactly on top of the original one.

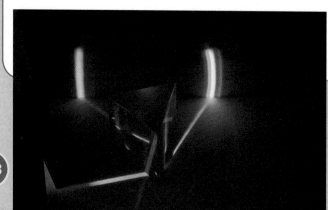

Isaac Newton

When Isaac Newton was 23 years old and in his fourth year at Cambridge University near London, the Great Plague struck, killing thousands of Londoners.

London's mayor wisely closed all of the schools and students were sent home!

Isaac returned to his mother's farm, where he spent eighteen months investigating gravity, developing a new system of maths called calculus, and explaining the physics of motion (just like most young men!).

He also turned his attention to explaining light. He held a three-sided piece of glass called a **prism**, so that a beam of sunlight could shine through it.

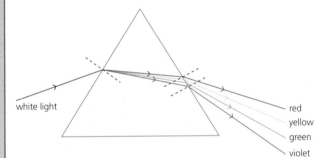

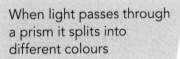

When light passes through a prism it splits into different colours

He found that the sunlight was refracted and changed direction as it passed through the glass, and it separated into different colours. These colours were the same as the colours of the rainbow.

This range of colours is known as the **visible spectrum** and many text books describe it by the seven colours:

red, orange, yellow, green, blue, indigo, violet.

The visible spectrum lies between infrared and ultraviolet radiation

It is difficult to distinguish between one colour and another and impossible really to count the actual number of colours. Isaac Newton found it convenient to describe the spectrum by the seven colours already listed and his description is still used today.

You may wonder why the prism, which is one material with one refractive index, should produce different ray paths for each colour of light?

The answer to this question caused a lot of debate in its time but most people now agree with the following explanation.

$\Rightarrow$

Isaac Newton

A ray of red light is monochromatic. That means the light ray consists of **one** colour (red!).

However, sunlight is not monochromatic.

It consists of a range of colours, and each of these colours is refracted by the glass prism by a different amount.

The prism therefore causes the light to spread out, or **disperse**, and a continuous visible spectrum is formed.

Red light waves which have long wavelengths of approximately 700 nm are refracted least, while violet light waves which have shorter wavelengths of approximately 400 nm are refracted most.

Gem secrets

Diamonds are sometimes referred to as having a 'fire' inside them.

This 'fire' or 'brilliance' is related to the amount of dispersion and refraction a diamond can produce.

Diamonds which produce a large angle between refracted red and blue light, are more brilliant, and sparkle more.

Of all the **transparent** substances on earth, diamond slows light by the greatest amount. Light is refracted most by dense materials because the light energy interacts with the material's structure as it passes through.

Diamond has a very large refractive index (in the region of 2.4) due to its unusual atomic structure and where the electrons are located.

Diamond is formed from **graphite** under intense pressure and temperature.

This formation process produces a high electron density which produces the large angle of light dispersion, which is seen as brilliance.

Diamond also reflects a lot of light, especially for a transparent material.

This ability to reflect light is called lustre and a diamond's lustre is forever!

Diamonds are cut and polished in such a way as to make the lustre and dispersion more obvious and easily seen. This unusual look and sparkle is what makes them desirable.

GLOSSARY

Angle of incidence the angle at which light strikes an object measured from the normal

Disperse to spread out

Graphite a form of the element carbon which is soft and black in appearance. It is found in the lead of pencils and also conducts electricity

Lenses a lens is an optical device made of glass or Perspex which helps our eyes to focus on near or distant objects

Normal an imaginary line used in calculating light passing through different media

Refraction the change in frequency of a wave when passing through different media

Transparent material which allows light to pass through

Visible spectrum the range of colours created when white light is separated

FORCES, ELECTRICITY AND WAVES

Waves

15

Lenses and magnification

Level 2 — What came before?

● SCN 2–11b

By exploring reflections, the formation of shadows and the mixing of coloured lights, I can use my knowledge of the properties of light to show how it can be used in a creative way.

Level 3 — What is this chapter about?

● SCN 3–11a

By exploring the refraction of light when passed through different materials, lenses and prisms, I can explain how light can be used in a variety of applications.

Lenses and magnification

The action of lenses

Optical lenses get their name from the Latin word for lentil!

Green and red lentils are vegetables and are typically used to make soup!

The lentil has a distinctive shape. It is thicker in the centre than at its edges.

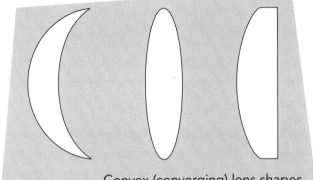

Convex (converging) lens shapes

Lenses shaped like the lentil are known as **convex lenses**.

When light enters such a lens its path is changed and the light **converges**. The point where the converging light is most concentrated is called the **focus** of the **lens**.

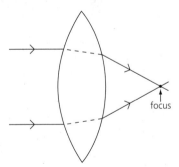

A lens which is thicker at the edges than the centre however is called a **concave lens**.

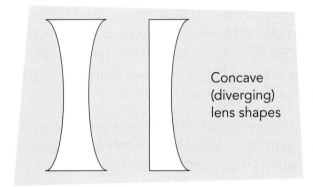

Concave (diverging) lens shapes

Concave lenses behave in the opposite fashion to convex lenses.

With a concave lens light is seen to **diverge**, or spread out. No focus can be achieved.

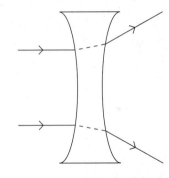

QUESTIONS

1 Using two prisms, can you arrange them to behave first like a convex (converging) lens, and second, like a concave (diverging) lens?

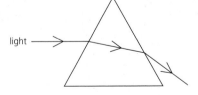

2 How do you think the amount of light deviation will vary with the thickness of the lens?

The appliance of science

One of the most important applications of lenses is in correcting eyesight.

Optometry is the branch of science dealing with vision and eye-care.

When light enters the eye from a distant object, the light rays are **parallel** to one another. Each eye of a person with normal eyesight refracts these light rays and brings them to a focus on the retina at the back of the eye.

When looking at a nearby object, such as a book or computer screen, the light rays entering the eye are **diverging**. A person with normal eyesight has no problem in refracting these diverging rays on to the retina of each eye.

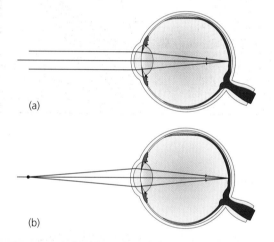

(a)

(b)

This is how a perfect eye views a distant object (a) and a nearby object (b)

Most refraction in the human eye is performed by the front surface, the *cornea* of the eye.

The cornea cannot be repaired easily if damaged, and this is why you should always wear safety goggles when carrying out certain experiments in science at school!

The convex lens inside the eye then makes fine focusing adjustments.

Muscles inside the eye can adjust the shape of the lens so that it focuses correctly!

The convex lens in the eye can be squeezed into a 'fatter', more rounded shape by these surrounding muscles which are called the *ciliary muscles*. This is perfect for focussing on close objects, such as the print in a book (like this!).

When looking at distant objects however, the eye lens is pulled 'thinner' and becomes less rounded. This is necessary for focussing on objects which are far away.

The ability of the eye to focus on objects which are distant or close up is called **accommodation**.

Light entering the eye through the cornea and lens is brought to a focus on the **retina** at the back of the eye.

The 'image' on the retina affects *receptors* which convert the image into electrical signals. These signals are then sent to the brain for processing.

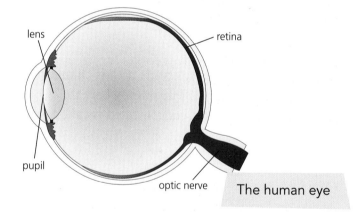

lens

retina

pupil

optic nerve

The human eye

Not everyone has perfect eyesight.

Teenagers typically have difficulty in reading a slightly distant board in the classroom, but they have much better short sight. We say they are **short-sighted** or **myopic**.

Middle-aged people typically have difficulty in reading computer screens and books close up, but have much better long sight. We say they are **long-sighted** or **hypermetropic**.

The appliance of science

If your eyes are not capable of normal accommodation then your optician needs to establish if you are short-sighted or long-sighted. This allows the optician to prescribe the correct shape and curvature (roundness) of lenses for spectacles.

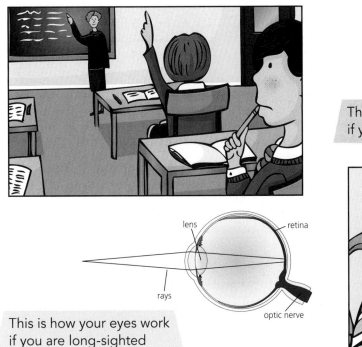

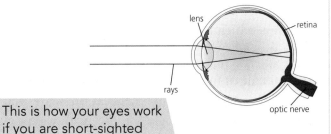

This is how your eyes work if you are short-sighted

This is how your eyes work if you are long-sighted

Reading glasses with convex lenses can be bought for a few pounds in supermarkets and can correct a middle-aged person's condition. However, glasses with concave lenses to correct a teenager's condition require dispensing by qualified staff.

Contact lenses

Contact lenses can also be used to correct eyesight defects and they have significant advantages over spectacles. They do not restrict the wearer's field of view, and the size of objects appears to be more natural. Contact lenses are widely used and they are particularly useful for someone who cannot wear normal spectacles because of abnormal corneas. They are also useful on the sports field where conventional spectacles might be damaged.

'Smart' contact lenses are currently being developed which incorporate sensors, solar cells, antennae, and radio chips. As well as correcting eyesight defects these lenses might eventually be able to transmit what a person is seeing. Can you think of some advantages and disadvantages of such a possibility?

QUESTIONS

1 From your knowledge of refraction, what shape of lens type placed in front of a short-sighted teenager's eye will push the focus 'backwards' and on to the retina?

2 What shape of lens placed in front of a long-sighted middle-aged person's eye will bring the focus forwards and on to the retina?

Refraction to reflection

When a light ray passes from an optically dense material (such as glass) to an optically less dense material (such as air), it speeds up and is refracted away from the **normal**.

As the angle of incidence in the dense material increases, there will come a point at which the light is no longer refracted at the material boundary, but is instead reflected. This is called **total internal reflection**.

As the angle of incidence in the dense material increases, the angle of refraction (in the less dense material) increases too. There comes a point however (with a large angle of incidence) when the light is no longer **refracted** at the boundary, but is **reflected** instead. This is called **total internal reflection** and it is illustrated in the following set of diagrams.

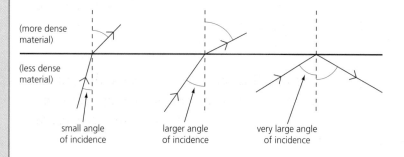

(more dense material)

(less dense material)

small angle of incidence   larger angle of incidence   very large angle of incidence

Light can actually be trapped in a material by continual total internal reflection. This is the principle of the **optical fibre** which is used in communications such as the world-wide web.

The angle of incidence, at which total internal reflection occurs with no refraction evident, is unique for each material.

It is known as the **critical angle** for that material.

Fibre optics

In *digital communications* a thin fibre of very pure optically dense glass is surrounded by a cladding glass, which is optically less dense.

Digital light signals can then be sent down the inner fibre, which is of the thickness of a human hair.

The same principle of total internal reflection is used too in medical **endoscopes** used for key-hole surgery and medical investigations.

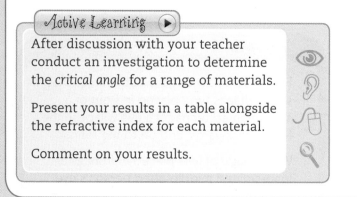

Active Learning ▶

After discussion with your teacher conduct an investigation to determine the *critical angle* for a range of materials.

Present your results in a table alongside the refractive index for each material.

Comment on your results.

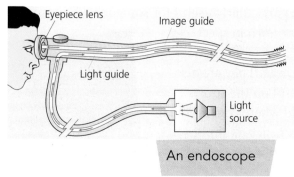

Eyepiece lens   Image guide

Light guide

Light source

An endoscope

Magnifying what we see

The microscope

You will already have studied the structure of cells using a laboratory microscope during the biological systems topic. You may even know the main parts of the microscope itself. What you perhaps do not understand yet, is how lenses can make objects look bigger!

The Dutch linen draper Antoni van Leeuwenhoek began grinding his own lenses in secret in Delft in 1671.

He wanted to be able to examine the quality of fabrics before purchase. He knew that if he held an object behind a round glass vase full of water, and looked through the vase, the object would appear magnified.

He reasoned that a very small glass ball would give much higher magnification.

Using a sheet of brass the size of a playing card, he drilled a small hole at the middle. In this hole he mounted a tiny glass lens, and this was his microscope.

The material (**object**) being studied was placed close to the lens and he moved his eye close to the other side of the lens, moving back and forth until he could see a magnified image of his object. What he saw magnified was very little. His field of view was restricted and it was very hard work on the eyes!

Antoni Van Leewenhoek's microscope

Modern day microscopes have more than one lens and are called compound microscopes.

In a compound microscope the magnification is done in two stages.

An **objective lens** forms a magnified real image at the end of a tube which is typically 160 mm long. This is shown in the following diagram.

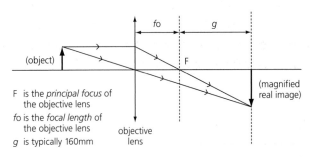

F is the *principal focus* of the objective lens
fo is the *focal length* of the objective lens
g is typically 160mm

The magnified real image formed at the end of the tube is then magnified further by a convex **eyepiece lens**. This is not shown in the diagram.

QUESTIONS

The ratio $\frac{g}{fo}$ is usually engraved on the microscope's objective casing, usually in a colour.

What number would you find on a standard length casing if fo is 4mm?

Active Learning ▶

The useful magnification of any microscope is ultimately determined by the nature of visible light itself! You simply cannot see anything smaller than the wavelength of light used.

Can you find out how scientists got round this limitation of the compound light microscope?

The telescope

A microscope allows us to see a tiny object by making it appear larger. If however we were to point a microscope at say a planet or a star to see it in more detail, we would have no success. Instead we use a **telescope**. The purpose of the telescope is not so much to make a distant object look larger, as to make it appear closer.

An astronomical telescope is an optical system consisting of two converging lenses.

The first lens directed towards the object, is called the **objective** lens. It has a long focal length and therefore looks fairly flat.

The objective lens forms a reduced, real image of the distant object. This image is however **inverted** (upside down).

The second lens has a short focal length and is far more curved. It is called the **eyepiece** lens. This lens acts as a magnifying glass forming an enlarged virtual image, which is still inverted.

Astronomers do not find the inverted image a problem and these telescopes are called **astronomical refracting telescopes**.

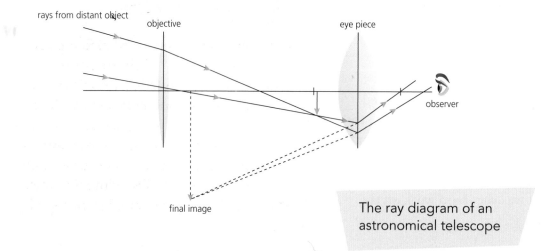

rays from distant object — objective — eye piece — observer — final image

The ray diagram of an astronomical telescope

There are other designs of telescopes which offer advantages over the refracting telescope. You should research the *Newtonian* and *Cassegrain* telescopes.

How do they work and what advantages do they have over the refracting telescope?

Binoculars

Binoculars are just a pair of telescopes mounted side by side.

However as they are used for viewing objects on Earth (*terrestrial* use), the images produced have to be turned the right way up! Two right-angled prims in each table are used for this correction, using the principle of total internal reflection. The prims act as **retro-reflectors**, sending light back in the direction it came from. Prisms make the instrument much more compact, because they effectively 'fold' up the light path.

Light from the objective lens is inverted left/right first by a horizontally placed right-angled prism, before reaching a second vertically mounted right-angled prism. In the second prism an up/down inversion takes place. The 'corrected' image is then viewed through the magnifying eyepiece lens. The diagram shows the up/down inversion.

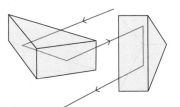

The combined action of the two prisms is represented in the following diagram.

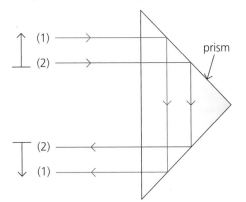

You should investigate retro-reflection further.

What applications and products have been developed that use retro-reflection?

The cat's eyes that mark the centreline of roads and the exits and entrances on motorways is a British invention now used the world over.

In 1934, a Yorkshire man named Percy Shaw patented this invention before establishing a company to manufacture cat's eyes in Halifax. The years of blackout during World War Two (1939–1945) accelerated the use of cat's eyes throughout the UK's road system.

How do cat's eyes work?

Shaw's original design consisted of four glass beads placed in two pairs facing opposite directions, embedded within a flexible rubber dome. The dome was mounted on a cast iron base which was buried in the road and fixed into position with tar. When vehicles drove over the rubber dome, the glass beads were pushed into the cast iron base which filled with rain water. The glass beads received a wash before popping up again!

The glass beads used in the cat's eyes have a convex front lens which refracts the light onto a spherical concave silvered surface at the rear, where reflection takes place. The reflected ray is parallel to the incoming ray.

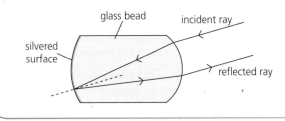

glass bead
incident ray
silvered surface
reflected ray

Refraction and rainbows

Rainbows can be seen in the sky when there are water droplets in the air from recent rain and the sun is shining from behind you at a low angle to the ground. Usually the bad weather is moving away from you and there are clearer skies behind you.

Sunlight from behind you enters the spherical water droplets in the clouds in front of you.

Inside the droplet the light is refracted before being reflected from the rear of the rain drop. The reflected light is refracted again as it leaves the water droplet in the direction of your eye.

Red light is refracted least on leaving the water droplet, so it is seen on the outside of the rainbow. Violet light is refracted most, so it is seen in the inside of the rainbow.

Only light from some raindrops will enter your eye at any particular time: red light from one drop, orange from another, blue from a third and so on. The overall effect is a rainbow which moves according to your position and the position of the sun.

Light reflecting and refracting in a water droplet

Refraction and meteorology

TV weather forecasters can predict general changes in the weather up to about a week in advance using computer models of the atmosphere, which run continually on supercomputers. However the models are not perfect. The physics of the atmosphere is not fully understood yet and models which try to predict too far into the future drift from reality. However UK Met. Office statistics show that the accuracy of forecasting has improved, and a current 3-day forecast is now as accurate as a 24-hour forecast was 20 years ago!

Accurate weather forecasting is important to keep the public safe. For example, sudden thunderstorms and lightning can create downdraughts of air which might affect air travel, or produce flash floods.

The accuracy of forecasting is improved by feeding real data on the true state of the atmosphere into the computer models. This process of assimilation brings the model closer to reality.

The position of a thunderstorm is determined by the amount of water vapour (humidity) distributed horizontally in the air. This water vapour condenses to form cloud droplets. The energy lost during the condensation of the water then heats the surrounding atmosphere and strong convection currents are set up, producing the characteristic stacked towers of thunderclouds.

cooler air

warm air

Refraction and meteorology

Thunderstorms are difficult to predict using computer models without a number of real humidity measurements being assimilated into the model.

Radar (radio detection and ranging) can determine the amount and location of existing rain. Electromagnetic radiation of 5cm wavelength is sent out by dish aerials. If more drops of rain are present in the atmosphere, more of the electromagnetic beam is reflected. The position of the rain shower can be calculated from the return time of the beam. If these individual measurements are then taken at different angles, a horizontal picture can be built up showing rain intensity and position.

This traditional method of mapping humidity is now being supplemented by a process called **radar refractivity**.

In the Earth's lower atmosphere, electromagnetic waves travel at about 99.97% of the speed of light in vacuum. However, on entering water they travel at only 75% of the speed of light. Electromagnetic waves like radar are slowed down by cold dense air, and high humidity air, which is laden with water.

Moist air increases the time taken for a radar wave to return from a fixed target in the surrounding landscape.

When the reflected radar wave is detected, it is found to be out of step with the transmitted wave. The amount the waves are out of step is then used to calculate the amount of 'slowing down' that has taken place during the journey to and from a fixed target. This increase in refractivity (refractive index) is a measure of the increase in the atmosphere humidity along this path.

The radar refractivity method is very sensitive and when information is gathered from a variety of directions and angles, a humidity map can be generated surrounding a particular radar dish. This map can then be used to detect the small changes in humidity that lead to localised and sudden, summer thunderstorms.

GLOSSARY

Concave lens an optical device made of glass or Perspex which makes rays spread out

Converge to come to a point

Convex lens an optical device made of glass or Perspex which makes rays come to a point

Diverge to spread out

Endoscope device for looking inside a person's body

Focus the point where light from a lens meets

Inverted upside down

Object the thing you are looking at

Optical fibre glass or Perspex fibre used to transmit light

Optometry the study and treatment of vision and related health issues

Parallel lines which will never meet

Retina an organ at the back of your eye

Telescope a device for looking at distant objects

Total internal reflection the property of a material which reflects light internally

PLANET EARTH

Space

Are we alone in the universe?

Level 2   What came before?

● SCN 2–06a

By observing and researching features of our solar system, I can use simple models to communicate my understanding of size, scale, time and relative motion within it.

Level 3   What is this chapter about?

● SCN 3–06a

By using my knowledge of our solar system and the basic needs of living things, I can produce a reasoned argument on the likelihood of life existing elsewhere in the universe.

Are we alone in the universe?

For many years people have considered the possibility that there is life beyond our own planet. Some believe it to be a fact and claim they have been abducted! This is especially true near Bonnybridge which is Scotland's 'hot-spot' for UFO hunters.

Many **astronomers** and **philosophers** have debated the issue that it is very unlikely for us to be the only form of life in the **universe**.

Initially people looked at our near neighbours such as the Moon, Mars and Venus as possibly having some form of life that we could perhaps observe or communicate with.

A number of old civilisations looked at the Moon and saw faces or structure in the way its craters are lit in its phases. A few had given god-like properties to the Moon and prayed to it to assist their lives down here.

More worryingly there are a number of people who believe that there are alien stations on the Moon (on the far side) and that NASA has covered this up to prevent the public knowing the truth. There are many books published and websites written that claim to provide evidence that this is the case. All are untrue and rely upon people's ignorance and belief that there something out there which is beyond scientific knowledge.

In the 1800's the Italian astronomer Giovanni Schiaparelli was looking at Mars with a powerful telescope and noticed channels or *Canalis* which may not have been caused naturally.

The American astronomer Percival Lowell looked at these closely and argued that they were the remnants of a lost Martian civilization, and that in order for them to be of such a size they must have been created by an advanced and much superior species.

Regrettably this was found to be untrue when looked at later with better telescopes.

The canals were introduced there, but they were caused by atmospheric processes, not by a great civilisation! Lowell saw something which was unusual and made up an explanation which quite wrong.

H. G. Wells was believed to have used this idea when he wrote his famous book about Martians invading the Earth called *The War of the Worlds*.

Before we can search for extra-terrestrial life we need to agree on what can be regarded as a living object.

Obviously people, animals, plants and insects etc. are clearly 'living' but is it acceptable to allow microscopically small things such as bacteria, viruses and single-celled organisms to be considered too?

On Earth, algae and microbes can exist in extremely harsh conditions. Some organisms have been found deep underwater at temperatures over 150°C! They don't even need sunlight as they can adapt the chemicals around them for nutrients.

What can living things do?

Biologists consider the following as the main characteristics of living **organisms**:

- Metabolism – adapting nutrients to feed.

- Growth – going through a life cycle.

- Reaction to stimuli – responding to conditions to improve their chances of survival.

- Reproduction – to produce more of their kind.

- Mutation – to change depending on their surroundings.

You will have noticed that intelligence or communication has not been mentioned. You don't need to communicate or be intelligent in order to be alive!

To begin to search for life outside our own planet, we need to narrow down what we wish to look at. If there are extreme temperatures say on a planet which do not allow life as we are aware of it, then we should not waste our efforts looking there.

For example, the huge gravitational attraction on Jupiter and its distance from the Sun means that it has a very harsh atmospheric environment which would make it very difficult for life, even in a very basic form, to exist.

Neptune or Pluto are not able to support life because they are so far from the Sun that the temperature can be below –200°C! Nothing we know of could survive in these conditions.

On Mercury the surface temperature can reach as high as 430°C and as low as –180°C.

Active Learning ▶

Activities

1 Is it worthwhile for us to search for some sort bacteria on another planet or moon? Give two reasons why you think this is a good idea or a bad idea.

2 Obtain a data booklet or use the Internet and indicate the maximum and minimum surface temperatures on:

   a) Earth

   b) Venus

   c) Mars

   d) Jupiter

3 Consider the possibility that we did find some bacteria at the North Pole of Mars.

   Should we bring it back to Earth for study?

   Justify your answer.

Again, life would not be able to exist.

To search for possible life we need to have a reasonable idea of what we are looking for.

There are billions of planets in space and if we just investigated them randomly we could spend a few million years without any hope of success!

It would seem reasonable to look at how we think life evolved on our planet and see what can be learned from that.

Numeracy $+ - \div \times$

1 Here is some numerical data on planets close to the Sun.

To get a better indication of how far the planets are from the Sun scientists use another unit. It is called an **astronomical unit** and means that the distance from the Sun to the Earth is 1 AU.

To calculate the distance in AU we divide the distance between that planet and the Sun by the distance between the Earth and the Sun. From the following table, the distance in AU for Mercury is therefore:

$$\frac{57\,910\,000}{149\,600\,000} = 0.39$$

Complete the table which follows and calculate the distance from the Sun to the other planets in astronomical units.

Planet	Distance from Sun (000 km)	Time for one revolution (days)	Distance in AU
Mercury	57 910	88	0.39
Venus	108 200	225	
Earth	149 600	365	1.0
Mars	227 940	687	
Jupiter	778 330	4 333	
Saturn	1 429 400	10 759	

Now use the table to calculate how many years it takes Jupiter and Saturn to orbit the Sun.

A short history of the Earth

It is believed that the Earth was formed some 4 600 000 000 years ago as the third planet in a series orbiting a star, our Sun.

It is also believed that all objects in our Solar System are made from the debris of stars that had come to the end of their existence. We are all made from Stars! You are a Star!

Early on in its formation, Earth was a fairly uniform rock with an early atmosphere made up mainly of hydrogen. The interior of the rock heated up and a molten core was formed. This heat energy was generated mainly by the radioactive decay of the materials that made up the rock-like planet.

The heated core then released gases from the planet's interior and these combined with the atmosphere that was already in existence. This newer atmosphere now contained water (H_2O), carbon dioxide (CO_2), methane (CH_4) and ammonia (NH_4).

Somehow this group of chemicals is thought to have combined to create the first set of chemicals needed for life to exist and ultimately to grow and adapt.

Another viewpoint is that while all the necessary chemicals were in place there is no real explanation as to exactly what created the first 'living things'.

Some scientists proposed that the first microbes or organisms came to Earth as part of some space debris or meteor collision and that this is a possible cause of the origins of life.

These explanations are not facts however; they are proposals by scientists that suggest that it is possible that life started this or that way. There are other theories too, but the one thing that they all have in common is that none of them are easy or straightforward to understand. Many scientists have also done experiments which show that certain chemicals can be produced under the right conditions, but no one has taken the raw chemicals and created a micro-organism yet!

$\Rightarrow$

A short history of the Earth

This subject is very contentious. There are strong scientific arguments which support some of the biochemistry and there are many philosophical and theological ideas which challenge the accepted science.

Regardless of what is the correct mechanism by which life was formed, we do have fossil evidence that shows bacteria in rocks which are 3 500 000 000 years old. This means that some form of life had evolved just over 1 100 000 000 years after the Earth was formed.

The following table makes it easier to understand our planet's timescale.

Millions of years ago	What happened
4600	origin of our planet
3900	oldest known rocks
3500	first evidence of life
510	oldest fossil fish
458	first land plants
375	amphibians appear
200	first mammals
160	first birds
65	dinosaurs extinct
3·4	fossils of human types
0·6	appearance of homo erectus (early man)
0·00003	Scotland last qualified for the world cup!

An interesting aspect of these numbers is how early some form of life was established and how long it took for that to evolve into some form of humanoid life.

What do we look for?

If we think there is other intelligent life out there in space, we should perhaps look for radio waves or something similar. However we ourselves have only been transmitting radio waves since after 1910 so that would narrow down our window a fair bit.

A better way might be to search for *Biomarkers* which are signs that there is some form of life on a planet. The European Space Agency has sent a satellite called *Darwin* into space for such a mission.

Such a satellite might examine the light coming from really distant planets. The light reflected by these planets can give us an indication of the gases in their atmospheres and hopefully this might suggest a planet worth investigating more closely.

QUESTIONS

Using the table above, answer the following questions.

1 When was planet Earth formed?

2 Which were the first kind of 'animals' to appear?

3 How long ago did dinosaurs became extinct?

4 How long has 'life' of some sort existed on our planet?

5 When did early man first appear?

Life in the Goldilocks Zone – why Earth?

What is it about our planet Earth that has allowed us to **evolve**?

Again, this is a difficult question which has been the topic of a huge amount of philosophical and scientific discussion for many hundreds of years.

Earlier, this chapter discussed how and when we think life originated in all its various forms.

What is it about Earth's position and size in Solar System that allows the conditions necessary to support life?

Most scientists accept that we need a number of physical and chemical conditions to allow life to form.

These are:

1 Energy: sunlight or chemical.

2 Water: mainly in liquid form.

3 Temperature: a range of temperature that is not too extreme.

4 Gravity: The gravitational field around our planet has allowed the atmosphere to form and is not too strong so that it retards the growth of small organisms.

The gravity of other planets is also an important factor. Jupiter is a large planet with a very strong gravitational pull. This acts like a 'space vacuum cleaner', attracting lots of *space debris* to it and hopefully away from us. It reduces the possibility of large meteors hitting Earth!

5 Atmosphere: The gases around our planet retain and circulate the heat throughout the surface and crust of the planet. This reduces extremes of temperature which could lead to poor conditions for survival. The Earth's atmosphere and also its magnetic field protect the planet from harmful radiation from the Sun.

The habitable (or Goldilocks) zone

Of all the conditions required for some form of life we believe the requirement for water to exist in liquid form at some point is the most important.

This means that the temperature of the planet must be in the region from 0 °C to 100 °C although the atmospheric pressure can affect whether water exists as a liquid.

In order to keep a planet at this temperature several conditions must be met.

⇒

The habitable (or Goldilocks) zone

1 One of these is a *planetary atmosphere* which absorbs and retains enough of the energy from the Sun to keep it at a relatively constant temperature. We have this. It is often referred to as the *greenhouse effect* and it has a large significance today! Of all the energy which reaches our planet from the Sun we absorb about 61% and reflect about 39%.

We refer to this as our planet's **albedo** and Earth has an albedo value of 0.39.

Albedo values which are too high or too low mean that the planet can be unbearably cold or unbearably hot. Our albedo and our atmosphere combine to give an average surface temperature on Earth of around 15 °C.

2 Another major factor is our distance from the Sun. We have to be at a distance which supplies us with a 'correct' amount of energy. Not too close, not too distant.

This region around the Sun where the energy isn't too great or too little is often referred to as the *Habitable Zone* (or the Goldilocks Zone). If a planet is not in this region there is little chance of liquid water being available.

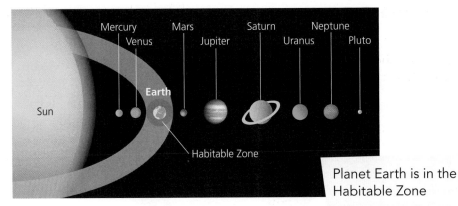

Planet Earth is in the Habitable Zone

Our planet is not only in the Habitable Zone as it exists today. It has been in this Habitable Zone for a long time! As the Sun formed and grew, the energy it gave out varied with time. The Habitable Zone has therefore changed too. Our planet was fortunate in that even as the Habitable Zone expanded, it remained in the right place. This means that the processes which led to the formation of life have operated over a very long time.

GLOSSARY

Albedo a measure of how much radiation our planet absorbs and reflects

Astronomer a person who studies space and the matter contained in it

Evolve to develop by evolution

Organism a living structure

Philosopher a person who tries to explain why things are the way they are

Universe all the known space

PLANET EARTH
Space

The search for other life

Level 2   What came before?

● SCN 2–06a

By observing and researching features of our solar system, I can use simple models to communicate my understanding of size, scale, time and relative motion within it.

Level 3   What is this chapter about?

● SCN 3–06a

By using my knowledge of our solar system and the basic needs of living things, I can produce a reasoned argument on the likelihood of life existing elsewhere in the universe.

The search for other life

There is no evidence at all to support any idea that there is any form of intelligent life in our **solar system**. No little green men or big red women for that matter!

Space probes are now travelling through our solar system and beyond, looking for various signs. Perhaps they will pick up some sign or indicator of life in the next few years. Space is so vast however it may be a long, long time before any signs (if any) are detected.

In our solar system however scientists are hopeful that there might be some primitive form of life in existence. The most likely locations are Mars, Europa and Titan.

Mars

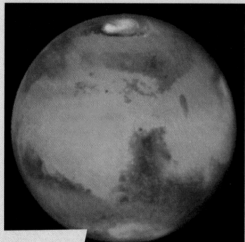

The planet Mars

This is the planet in our solar system which has most similarities with Earth so it would seem reasonable to consider first. It is probably too cold now but there may be some form of life near any warmer sections in the crust. In the past when it was warmer, liquid water may have formed and of course this is one of the essentials for life.

Mars has an atmosphere but it is much thinner than that of Earth so dangerous cosmic and UV radiation will be able to reach the surface.

Mars has a fairly rocky surface. These surfaces and outcrops allow the essentials for life to gather in certain areas, and so increase the possibility of life occurring. If the surface is a large gurgling liquid the different chemicals splashing around might rarely get the chance to meet.

Mars has North and South ice-capped Poles like Earth. They are much, much, colder however, and the 'ice' is not frozen water but frozen carbon dioxide. Perhaps deep under the surface water has frozen and is now trapped by the frozen carbon dioxide above.

Europa

Europa is one of Jupiter's moons

Europa is one of Jupiter's moons.

At its distance, it is a cold inhospitable body. The entire moon is covered by thick layer of ice.

The surface temperature is −170 °C and it has no atmosphere. It should in theory therefore have no possibility of life.

However, photographs taken by the *Galileo* space probe showed some features on the surface of the moon which were not expected! (They weren't footprints, regretfully.)

The search for other life

Photographs showed hundreds of lines crossing the icy surface, reminiscent of Arctic ice. Might it be possible that these lines are caused by floes moving and creaking against each other? If so, there is a possibility that liquid water might exist under the surface. This would be of great significance.

Europa is too cold is it not? Possibly!

However Europa will feel the force of gravitational attraction from Jupiter and all the other moons in the area. These 'tidal forces' pull the surface of Europa out of shape and can lead to spots where continual movement might lead to areas where the core could generate some heat.

This in turn might allow life to evolve.

Titan

Titan is Saturn's largest moon

Titan is Saturn's largest moon and the only moon in the solar system with a dense atmosphere.

It is possible that this atmosphere is very similar to that of Earth when Earth was younger and life was just starting to evolve.

It is a very cloudy moon, and has an orange colour. One hope is that this cloud conceals oceans or large quantities of chemicals in what is likely to be a fairly wild environment. There is a possibility that the conditions for life to evolve are there.

In 2005 a space probe landed on this moon. The probe, called *Huygens*, sent back many images of this mysterious moon as it fell through its atmosphere and landed on the surface. It had to send the images to another probe called the *Cassinni* probe and it in turn relayed them to Earth.

Initial findings are that its atmosphere like Earth's is full of nitrogen. It also has some methane and there are lakes or rivers of ethane on the surface. Such organic molecules are similar to those on Earth, so there is a possibility of life but again it is so far away from the Sun that it's surface temperature is −180 °C. There is probably no life now but perhaps in the past it was warm enough for some microbes to form.

An interesting point to consider is that in the future when our Sun expands and grows into a *Red Giant* it will push its habitable zone outwards. Life will not exist on Earth then but may form on Titan when it heats up.

Conclusion

What then is the chance of life on these moons and planets?

Very very low!

Even to discover such life would be very difficult and expensive and even if we did get close enough would we have the correct instruments or devices which will allow us to confirm the existence of some sort of life.

Could we bring it back?

Would we want to bring back some microbes from another planet?

The search for intelligent life

Is there intelligent life elsewhere? Not just life, but intelligent life?

This is a much more difficult question to answer because there are so many variables.

For example, there is the possibility that there may have been intelligent life somewhere but that the planet ran out of resources and the civilisation ended. Perhaps there is a similar life form to us on a planet somewhere in the furthest reaches of the Universe and we will never be aware of its existence.

There may also be a planet with life similar to that of our early humanoid or prehistoric ancestors. Its inhabitants are intelligent but do not yet have the technology to either detect us or be detected by us.

A reasonably scientific way of considering this issue was put forward by Dr Frank Drake in 1961. He was an astronomer working at the National Radio Astronomy Centre in West Virginia in America and wanted to explore the idea that we could estimate the possible number of civilisations we could communicate with.

Dr Frank Drake

He produced an equation, known as the *Drake equation* which is

$$N = R^\star\, f_p\, n_e\, f_l\, f_i\, f_c\, L$$

All terms on the right-hand side are multiplied.

In this equation:

- N = the number of communicative civilizations

- R^* = the rate of formation of suitable stars (such as our Sun)

- f_p = the fraction of those stars with planets

- n_e = the number of Earth-like worlds per planetary system

- f_l = the fraction of those Earth-like planets where life actually develops

- f_i = the fraction of life sites where **intelligence** develops

- f_c = the fraction of communicative planets (those on which electromagnetic communications technology develops)

- L = the 'lifetime' of communicating civilizations.

It's a complicated equation and looks quite daunting at first but it just lists the factors which should be considered when dealing with the possibility of advanced, intelligent life which is technologically capable.

The arguments that scientists and others have now are about deciding the numerical values to put into the equation and so calculate an answer!

Consider the question, 'what is the fraction of stars with planets?' What is a reasonable answer? Recent opinions appear to suggest between 20% and 50%! This is nothing more than reasoned guesswork and it could be out by a large amount.

Consider too 'what is the fraction of life sites where intelligent life develops?'

Answers ranging from 0 to 1 have been suggested which is as large a variation as you can have!

All of the factors have a large 'fudge' or unreliability built in.

Taking this into account gives us a quick range of answers! A recent serious attempt at the calculation came up with a value of 2.3. Drake's own estimate however is about 10 000.

Therefore taking into account all our recent discoveries and understanding of astrophysics and biology there are probably very few civilisations at present which we could contact. However this is much more speculation then serious scientific calculation!

SETI

What would be a reasonable way of trying to find intelligent life?

Obviously some form of astronomical search would be best.

Radio telescopes are probably the best way of trying to find out if there are other intelligent civilisations. If there are intelligent species they would have to develop radio communication before they could even travel in space.

It would be reasonable therefore to 'listen' for these radio waves as they spread through space.

$\Rightarrow$

SETI

As they spread the signals get very weak however, so we need to build large telescopes to detect them. The most recent one is the *Allen Telescope Array* which became operational in 2007. Over time the array will expand to 350 dishes.

The Allen Telescope Array at the University of California

It scans the Universe picking up radio waves which are travelling through space and tries to detect any pattern or repeated signal which would not be produced naturally by stars and galaxies.

This is an **incredibly** difficult task!

The **Search for Extraterrestrial Intelligence (SETI) Institute** is an organisation which co-ordinates work in this area. The Institute researches all areas of science which may impact on what we do to find evidence of intelligent life elsewhere in the Universe.

SETI has been analysing radio signals for over 30 years, looking for repeated patterns or codes or something which cannot be explained by natural phenomena.

So far there has been nothing but it doesn't stop people from looking. Professor Paul Davies has recently written a book called *The Eerie Silence* in which he describes the work being done by SETI.

Points to ponder

We have been emitting radio waves for about 100 hundred years.

Any civilisation which could detect us must be within the distance radio waves can travel in this time. This narrows down the numbers of stars with planets supporting intelligent life with whom we could communicate.

The main reason we have not heard anything of note from ETs yet is that our searches so far have been concentrated in a relatively tiny area of space around us.

Space is simply enormous and the distances which signals have to travel is difficult to imagine. Such distances can be tens of thousands of light-years.

Drake spoke to the Royal Society in January 2010 and said:

'In searching for **extraterrestrial** life, we are both guided and hindered by our own experience. We have to use ourselves as a model for what a technological civilisation must be, and this gives us guidance for what technologies might be present in the Universe.

At the same time, this limits us because we are well aware that all the technologies that might be invented have not been invented; and in using ourselves as a model, we may not be paying attention to alternatives, as yet undiscovered and as yet unappreciated by us.'

⇨

This means we may be looking for the wrong sort of signal. Maybe radio waves are not what we should be searching for!

In other words, we've been listening for extraterrestrials' radio signals but this may not be how they're trying to announce their presence.

It's one of the reasons why SETI, in the last few years has also started to look for optical flashes that might originate from powerful alien lasers systems.

Drake also mentioned that modern technology is possibly making us more invisible to the extraterrestrials who might be searching for us.

The signals from Earth most likely to reach distant civilisations are our TV broadcasts.

But the switchover from analogue to digital television means 'our voice' is being diminished.

Part of this is down to the TV satellites which deliver targeted beams to the Earth's surface, and also to cable TV which runs direct to the home underground. Both don't 'bleed' as much radiation into space as the older high-power analogue TV transmitters. Also the digital signal itself is more difficult to interpret.

Drake says this may mean that in future we have to establish a dedicated system of beacons to broadcast our presence.

'There are people who are saying we should be running a beacon – a simple message that we send to one star after another, pointing out that we exist.

When you think about that, you quickly reach the conclusion that there should be two beacons – one that's easy to detect and has only the information on it that tells you what radio frequency you should tune to to get the other beacon with a great deal more information.

Right now, we don't have to do this because we're sending all this information through our television, but when the Earth goes quiet it makes much more sense.'

Such a search is fascinating. Does it really have any point, and what would we do if we got a contact?

Would we be filled with fear or excitement?

How should we initiate a dialogue, knowing a reply might not come back for hundreds of years because of the travel time between our two locations?

Or would shocked and scared Earthlings immediately want to 'hang up' as if to say 'sorry, wrong number'?

GLOSSARY

Extraterrestrial beyond the planet Earth

Intelligence property of an organism to think, analyse and communicate

SETI Search for Extraterrestrial Intelligence

Solar system a collection of planets which orbit a star

Curriculum for Excellence mapping grid

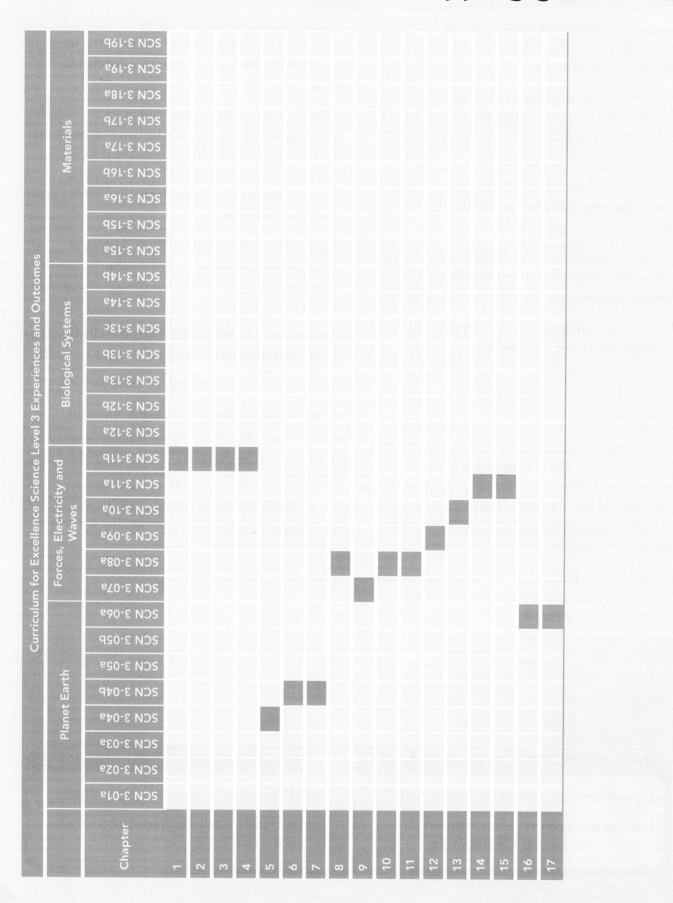

The Publishers would like to thank the following for permission to reproduce copyright material:

Photo credits
p.7 (top left) © bilwissedition Ltd. & Co. KG/Alamy, (top right) Hulton Archive/Getty Images, (bottom) Getty Images/The Bridgeman Art Library; p.8 (top) Getty Images/The Bridgeman Art Library, (middle) © World History Archive/Alamy, (bottom) Gary Doak/Scottish Viewpoint; p.10 © Russell Illig/Photodisc/Getty Images; p.11 (top all) © 1997 Tony Gable and C Squared Studios/Photodisc/Getty Images; p.12 (top) PAUL RAPSON/SCIENCE PHOTO LIBRARY, (bottom) GIPHOTOSTOCK/ SCIENCE PHOTO; p.17 (top) JPL/NASA, (bottom) Max Alexander; p.19 (top) © Mark A. Johnson/Alamy, (middle) DR P. MARAZZI/ SCIENCE PHOTO, (bottom) ANTONIA REEVE/SCIENCE PHOTO; p.20 (bottom) Wellcome Library, London. Wellcome Images; p.21 © Andersen Ross/cultura/Corbis; p.23 Mike Goldwater/Getty Images; p.24 Mike Harrington/Digital Vision/Getty Images; p.25 (bottom left, bottom middle) BJORN RORSLETT/SCIENCE PHOTO LIBRARY, (bottom right) Martyn F. Chillmaid; p.26 (bottom) Courtesy of SOHO/ Extreme Ultraviolet Imaging Telescope (EIT) consortium. SOHO is a project of international cooperation between ESA and NASA; p.27 NASA/ JPL-Caltech/POS S-II/DSS; p.29 © INTERFOTO/Alamy; p.30 © David Hoffman Photo Library/Alamy; p.34 (top left) X-ray: NASA/CXC/CfA/R. Kraft et al; Radio: NSF/VLA/Univ.Hertfordshire/M.Hardcastle; Optical: ESO/VLT/ISAAC/M.Rejk uba et al., (middle right) DR SETH SHOSTAK/ SCIENCE PHOTO LIBRARY, (bottom left) © Warren Kovach/Alamy; p.35 © Luis Carlos Torres/istockphoto.com; p.39 (top middle, top right) Ingram Publishing, (middle) © ephotocorp/Alamy, (bottom) Wellcome Images; p.40 (middle left) DR R. CLARK & M.R. GOFF/SCIENCE PHOTO LIBRARY, (bottom left) NRSC LTD/SCIENCE PHOTO LIBRARY, (bottom right) NASA/JPL-Caltech/C. Engelbracht (University of Arizona); p.42 © imagebroker/Alamy; p.43 (top right, middle right) Martyn F. Chillmaid, (bottom left) © Alex Segre/Alamy, (bottom right) © Imagentix/Alamy; p.44 © Tim Fletcher; p.48 Martyn F. Chillmaid; p.55 © (top) keith morris/Alamy, (middle) © Mark Boulton/Alamy, (bottom) Jake Wyman/Photodisc/Getty Images; p.58 (top right) © Skyscan Photolibrary/Alamy, (bottom left) © Robert Harding Picture Library Ltd/ Alamy, (bottom right) Scott Barbour/Getty Images; p.60 ©RIA Novosti/ TopFoto; p.63 (top from all) Werran/Ochsner/Getty Images, © George Doyle/Stockbyte/Getty Images, (middle from left) © Emilio Ereza/Alamy, Bob Thomas/Getty Images, (bottom all) Martyn F. Chillmaid; p.64 (top) © Steven May/Alamy, (bottom left) © Motoring Picture Library/Alamy, (bottom right) Insurance Institute for Highway Safety, HO/AP Photo/Press Association; p.65 (left) Mercedes Benz/AP Photo/Press Association Images, (right) TRL LTD./SCIENCE PHOTO LIBRARY; p.66 (middle) © dave jepson/Alamy, (bottom) © Derek Croucher/Alamy; p.67 (top right) © Patrick Eden/Alamy, (middle left) Juan Silva/Photodisc/Getty Images, (middle right) EDWARD KINSMAN/SCIENCE PHOTO; p.68 Martyn F. Chillmaid; p.69 Sipa Press/Rex Features; p.72 (top) © Radius Images/ Alamy, (middle left) © Judith Collins/Alamy, (middle right) Ker Robertson/ Getty Images, (bottom) Martyn F. Chillmaid; p.73 (top) Andy Hooper/Daily Mail/Rex Features, (middle) © Andi Duff/Alamy, (bottom) © Joggie Botma/Alamy; p.74 (top) © Tina Manley/North America/Alamy, (middle) Rex Features, (bottom) Keith Meatheringham/Rex Features; p.76 (top) Giuliano Bevilacqua/Rex Features, (middle left) © StockTrek/Photodisc/ Getty Images, (middle right) © Oleksiy Maksymenko/Alamy, (bottom) © JR Tokai/Reuters/Corbis; p.77 (top right) © Ed Simons/Alamy, (bottom left) © Roy Childs/Alamy, (bottom right) Martyn F. Chillmaid; p.82 (middle right) NASA Goddard Space Flight Center, (bottom middle) Hulton Archive/Getty Images, (bottom right) © David J. Green - studio/ Alamy; p.83 (left) NASA Nov. 25, 2009, (right) SCIENCE PHOTO LIBRARY; p.84 (middle right) © Bob Richardson/Alamy; p.85 (top) © Graeme Peacock/Alamy, (bottom) © John McKenna/Alamy; p.86 (left) © Premier/Alamy, (right) © David Angel/Alamy; p.90 (top right) Martyn F. Chillmaid, (middle left) LAWRENCE MIGDALE/SCIENCE PHOTO LIBRARY, (middle right) Martyn F. Chillmaid, (bottom middle) © Marvin Dembinsky Photo Associates/Alamy, (bottom right) © D. Hurst/Alamy; p.91 (all) Martyn F. Chillmaid; p.92 (top middle) © 1997 Scott T. Baxter/ Photodisc/Getty Images, (top right) Hugh Sitton/Stone/Getty Images, (bottom left) Martyn F. Chillmaid, (bottom right) MIKE AGLIOLO/ SCIENCE PHOTO LIBRARY; p.94 Martyn F. Chillmaid; p.96 © Digital Vision/Getty Images; p.106 (top) © Picture Press/Alamy; p.107 (middle left) SHEILA TERRY/SCIENCE PHOTO LIBRARY; p.111 (top) © Andrew Butterton/Alamy, (middle) © Blue Room/Alamy, (bottom) © James Jackson/Alamy; p.112 (top) © Picture Contact/Alamy, (bottom) © fabioberti.it/Alamy; p.115 © Leslie Garland Picture Library/Alamy; p.120 PASIEKA/SCIENCE PHOTO LIBRARY; p.125 © Nick Rowe/ Photodisc/Getty Images; p.127 Courtesy of Wikimedia Commons; p.129 © Leslie Garland Picture Library/Alamy; p.130 (top left) © PhotoLink/ Photodisc/Getty Images, (middle right) © FogStock/Alamy, (bottom right) © blickwinkel/Alamy; p.131 © Ryan McGinnis/Alamy; p.134 (left) Seth Joel/Getty Images, (right) NASA/JPL/University of Arizona; p.135 © Wolfgang Polzer/Alamy; p.142 (left) © NASA/JPL- Caltech, (right) NASA/Goddard Space Flight Center Scientific Visualization Studio; p.143 NASA/JPL/University of Arizona; p.144 DR SETH SHOSTAK/SCIENCE PHOTO LIBRARY; p.146 DR SETH SHOSTAK/SCIENCE PHOTO LIBRARY. p.26 (top), p.49, p.82 (middle left), p.93 (all), p.98, p.99 (all), p.100, p.101 (top), p.102, p.103, p.106 (bottom), p.107 (middle right, bottom), p.109, p.118 © Paul Chambers. p.6, p.15, p.28, p.37, p.62, p.71, p.81, p.89, p.97, p.105, p.114, p.122 MEDICAL RF.COM/SCIENCE PHOTO LIBRARY.
p.46, p.51, p.57, p.133, p.141 © Digital Stock
p.11 (bottom), p.20 (middle), p.43 (bottom middle), p.84 (middle left, bottom left), p.101 (bottom) © Hodder Gibson.

Every effort has been made to trace all copyright holders, but if any have been inadvertently overlooked the Publishers will be pleased to make the necessary arrangements at the first opportunity.